Site layout planning
for daylight and sunlight:
a guide to good practice

PJ Littlefair

BRE

constructing the future

BRE is the UK's leading centre of expertise on building and construction, and the prevention and control of fire. Contact BRE for information about its services, or for technical advice, at:
BRE, Garston, Watford WD25 9XX
Tel: 01923 664000
Fax: 01923 664098
email: enquiries@bre.co.uk
www.bre.co.uk

BRE publications are available from:
www.ihsbrepress.com
or
IHS BRE Press
Willoughby Road
Bracknell RG12 8FB
Tel: 01344 328038
Fax: 01344 328005
brepress@ihsatp.com

BR 209
ISBN 1 86081 0411
Previously published by BRE under
ISBN 0 85125 506 X

Contents

How to use this guide

Before using this guide, read the introduction which sets out the scope and nature of the guidance.

New development

To design for good daylighting and sunlighting within a new development, use Sections 2.1 (Skylight) and 3.1 (Sunlight), and Appendix C. Section 4 explains how to plan for solar heat gain. If there is a conflict with other requirements, Section 5 gives advice.

Gardens and open spaces

Section 3.3. deals with sunlight in gardens and other open spaces between buildings.

Existing buildings

To protect the daylighting and sunlighting of existing buildings use Sections 2.2 (Sky-light), 3.2 (Sunlight), 5.8 (Solar dazzle), and Appendix E (Rights to light).

Adjoining development land

The daylighting of adjoining development land is covered in Section 2.3.

The other appendices contain calculation methods and data to help to assess the daylighting and sunlighting within a site layout.

1 Introduction

People expect good natural lighting in their homes and in a wide range of non-domestic buildings. Daylight makes an interior look more attractive and interesting as well as providing light to work or read by. Access to skylight and sunlight helps to make a building energy-efficient; effective daylighting will reduce the need for electric light, while winter solar gain can meet some of the heating requirements.

The quality and quantity of natural light in an interior depend on two main factors. The design of the interior environment is important: the size and position of windows, the depth and shape of rooms, the colours of internal surfaces. But the design of the external environment also plays a major role: whether obstructing buildings are so tall that they make adequate daylighting impossible, or whether they block sunlight for much of the year.

This guide gives advice on site layout planning to achieve good sunlighting and daylighting, within buildings and in the open spaces between them. It is intended to be used in conjunction with the interior daylight recommendations in the British Standard BS 8206:Part 2[1] and the *Applications manual: window design*[2] of the Chartered Institution of Building Services Engineers (CIBSE). It complements them by providing advice on the **planning of the external environment**. If these guidelines on site layout are followed, along with the detailed window design guidance in the British Standard[1] and CIBSE manual[2], there is the potential to achieve good daylighting in new buildings, to retain it in existing buildings nearby, and to protect the daylighting of adjoining land for future development.

Other sections give guidance on passive solar site layout, on the sunlighting of gardens and amenity areas, and briefly review issues like privacy, enclosure, microclimate, road layout and security. The appendices contain methods to quantify access to sunlight and daylight within a layout.

While this guide supersedes the 1971 Department of the Environment document *Sunlight and daylight* which is now withdrawn, the main aim is the same — to help to ensure good conditions in the local environment, considered broadly, with enough sunlight and daylight on or between buildings for good interior and exterior conditions.

The guide is intended for building designers and their clients, consultants and planning officials. The advice given here is not mandatory and this document should not be seen as an instrument of planning policy. Its aim is to help rather than constrain the designer. Although it gives numerical guidelines, these should be interpreted flexibly because natural lighting is only one of many factors in site layout design (see Section 5). In special circumstances the developer or planning authority may wish to use different target values. For example, in a historic city centre a higher degree of obstruction may be unavoidable if new developments are to match the height and proportions of existing buildings. Alternatively, with a building where natural light and solar gain are of special importance, less obstruction and more sunlight and daylight may be deemed necessary. The calculation methods in Appendices A, B and G are entirely flexible in this respect. Appendix F gives advice on how to develop a consistent set of target values for skylight in such circumstances and Appendix C shows how to relate these to interior daylighting requirements.

2 Light from the sky

2.1 New development

The quantity and quality of daylight inside a room will be impaired if obstructing buildings are large in relation to their distance away. The distribution of light in the room will be affected as well as the total amount received. The following guidelines may be used for houses and any non-domestic buildings where daylight is required.

At the site layout stage in design, window positions will often be undecided. So, for checking purposes, a series of reference points 2 m above ground level on each main face of the building may be used. The 2 m height corresponds to the top part of ground-floor windows; for buildings with basement windows, or those with no ground-floor windows but with windows higher up (for example flats mounted on a podium), a level of 2 m above the base of the lowest storey may be taken (Figure 1).

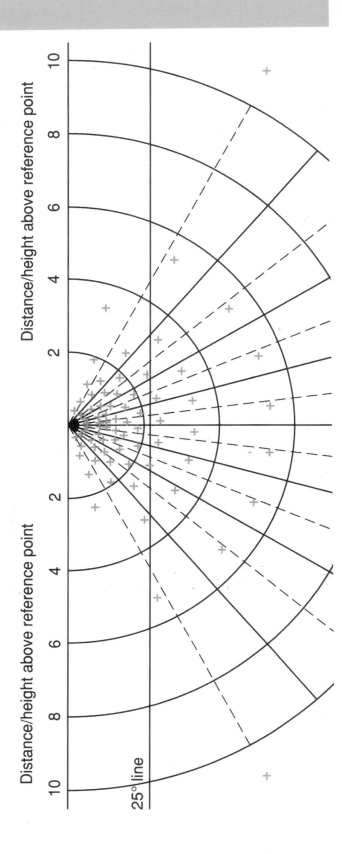

As a first check, draw a section in a plane perpendicular to each main face of the building (Figure 2). If none of the obstructing buildings subtends an angle to the horizontal (at the 2 m reference height) greater than 25°, then there will still be the potential for good daylighting in the interior (Figure 3).

Section

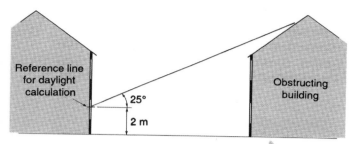

Figure 2 Section in plane perpendicular to the main face of the building

Section

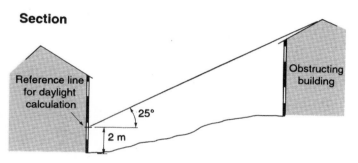

Figure 3 On sloping sites overshading is more of a problem and greater spacing is required to obtain the same access to daylight for buildings lower down the slope

If an obstructing building is taller than this, good daylighting may still be achievable, provided the obstruction is not continuous and is narrow enough to allow adequate daylight around its sides. The amount of skylight falling on a vertical wall or window can be quantified as the vertical sky component. This is the ratio of the direct sky illuminance falling on the vertical wall at a reference point, to the simultaneous horizontal illuminance under an unobstructed sky. The 'Standard Overcast Sky' of the Commission Internationale de l'Eclairage (CIE — International Commission on Illumination) is used and the ratio is usually expressed as a percentage. The maximum value is almost 40% for a completely unobstructed vertical wall. The vertical sky component on a window can be related to the average daylight factor in a room, which is one basis for the British Standard recommendations on interior daylighting (see Appendix C).

Vertical sky components may be calculated using the skylight indicator (Appendix A) or Waldram Diagram (Appendix B). Note that all obstructing buildings will have an effect, not just those on the same site. For calculation purposes, trees may be

ignored unless they form dense continuous belts. However, the siting of trees is important; where possible locate them away from windows.

For a room with non-continuous obstructions there is the potential for good daylighting provided that the vertical sky component, at the window position 2 m above ground, is not less than the value for a continuous obstruction of altitude 25°. This is equal to a vertical sky component of 27%.

If window locations are flexible, calculate the vertical sky component at a series of points on each main face of the building 2 m above the ground (or lowest storey base) and no more than 8 m apart. The building face as a whole should have good daylighting potential if every point on the 2 m high reference line is within 4 m (measured sideways) of a point with a vertical sky component of 27% or more.

Where the vertical sky component is found to change rapidly along a facade, it is worthwhile, if possible, to site windows where most daylight is available. The situation often occurs at the internal corners of courtyards or L-shaped blocks. If windows are sited close to these corners, levels of daylight will be poor and there may be a lack of privacy (Figure 4). The skylight indicator (Appendix A) or the '45° approach' suggested in Section 2.2 for domestic extensions, can be used to check how far from an internal corner windows need to be sited to receive enough light from the sky.

Figure 4 Rooms looking out from the internal corners of courtyards can often be gloomy and lack privacy

In some cases, for example with a standard house design, window positions may already be known. The vertical sky component can then be calculated at the centre of each window. In the case of a floor-to-ceiling window, such as a patio door, a point 2 m above ground on the centre line of the window may be used. Again, a vertical sky component of 27% or more indicates the potential for good daylighting. The interior daylighting of the building can then be checked easily using the method described in Appendix C.

Where space in a layout is restricted, interior daylighting may be improved in a number of ways. An obvious one is to increase window sizes. The best way to do this is to raise the window head height, because this will improve both the amount of daylight entering and its distribution within the room (Figure 5).

Improving external surface reflectances will also help. Light-coloured building materials and paving slabs on the ground may be used. However, maintenance of such surfaces should be planned to stop them

Figure 5 In Georgian streets the small spacing-to-height ratio is compensated for by tall windows. Note how window-head height increases for the lower floors which are more heavily obstructed

discolouring. Often the benefits will not be as great as envisaged, partly because of ageing of materials and partly for geometrical reasons. An obstructed vertical building surface will receive light from less than half the sky. Even if it is light coloured its brightness can never approach that of unobstructed sky.

Finally, one important way to plan for good interior daylight is to reduce building depth (window wall to window wall). Even on a totally unobstructed site there is a limit to how deep a room can be while remaining properly daylit. The presence of obstructions may reduce this limiting depth still further. Appendix C gives details of how to calculate these limiting room depths for good daylighting.

> ### Summary
> In general, a building will retain the potential for good interior diffuse daylighting provided that on all its main faces:
>
> (a) no obstruction, measured in a vertical section perpendicular to the main face, from a point 2 m above ground level, subtends an angle of more than 25° to the horizontal;
>
> or
>
> (b) if (a) is not satisfied, then all points on the main face on a line 2 m above ground level are within 4 m (measured sideways) of a point which has a vertical sky component of 27% or more.

2.2 Existing buildings

In designing a new development or extension to a building, it is important to safeguard the daylight to nearby buildings. A badly planned development may make adjoining properties and their gardens gloomy and unattractive, annoying their occupants and even, in some cases, infringing rights to light (see later in this Section). The guidelines given here are intended for use with adjoining dwellings and any existing non-domestic buildings where the occupants have a reasonable expectation of daylight; this would normally include schools, hospitals, hotels and hostels, small workshops and most offices. Gardens and open spaces are dealt with in Section 3.3.

Note that numerical values given here are purely advisory. Different criteria may be used, based on the requirements for daylighting in an area viewed against other site layout constraints.

A modified form of the procedure adopted for new buildings can be used to find out whether an existing building still receives enough skylight. First, draw a section in a plane perpendicular to each affected main window wall of the existing building (Figure 6). Measure the angle to the horizontal subtended by the new development at the level of the centre of the lowest window. If this angle is less than 25° for the whole of the development then it is unlikely to have a substantial effect on the diffuse skylight enjoyed by the existing building.

Section

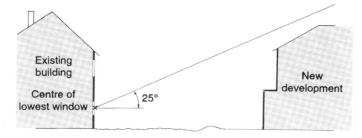

Figure 6 Section in plane perpendicular to the affected window wall

If, for any part of the new development, this angle is more than 25°, a more detailed check is needed to find the loss of skylight to the existing building. Both the total amount of skylight and its distribution within the building are important.

Any reduction in the total amount of skylight can be calculated by finding the vertical sky component at the centre of each main window. (In the case of a floor-to-ceiling window, such as a patio door, a point 2 m above ground on the centre line of the window may be used.) The reference point is in the external plane of the window wall. Windows to bathrooms, toilets, storeroom, circulation areas and garages need not be analysed. The vertical sky component can be found by using the skylight indicator (Appendix A) or Waldram Diagram (Appendix B).

If this vertical sky component is greater than 27% then enough skylight should still be reaching the window of the existing building. Any reduction below this level should be kept to a minimum. If the vertical sky component, with the new development in place, is both less than 27% and less than 0.8 times its former value, then occupants of the existing building will notice the reduction in the amount of skylight. The area lit by the window is likely to appear more gloomy, and electric lighting will be needed more of the time.

The impact on the daylighting distribution in the existing building can be found by plotting the no-sky line in each of the main rooms. For houses this would include living rooms, dining rooms and kitchens. Bedrooms should also be analysed, although they are less important. In non-domestic buildings each main room where daylight is expected should be investigated. The no-sky line divides points on the working plane which can and cannot see the sky. (In houses the working plane is assumed to be horizontal and 0.85 m high; in offices 0.7 m high; in special interiors like hospital wards and infant school

classrooms a different height may be appropriate.) Areas beyond the no-sky line, since they receive no direct daylight, usually look dark and gloomy compared with the rest of the room, however bright it is outside. According to the British Standard[1], supplementary electric lighting will be needed if a significant part of the working plane lies beyond the no-sky line. Appendix D gives hints on how to plot the no-sky line.

If, following construction of a new development, the no-sky line moves so that the area of the existing room which does receive direct skylight is reduced to less than 0.8 times its former value, then this will be noticeable to the occupants, and more of the room will appear poorly lit. This is also true if the no-sky line encroaches on key areas like kitchen sinks and worktops.

These guidelines need to be applied sensibly and flexibly. There is little point in designing tiny gaps in the roof lines of new development in order to safeguard no-sky lines in existing buildings. If an existing building contains rooms lit from one side only and greater than 5 m deep, then a greater movement of the no-sky line may be unavoidable. Another important issue is whether the existing building is itself a good neighbour, standing a reasonable distance from the boundary and taking no more than its fair share of light.

However, as a general rule the aim should be to minimise the impact to existing property. This is particularly important where successive extensions are planned to the same building. In this case the total impact on skylight of all the extensions needs to be calculated and compared with the guidelines.

For domestic extensions which adjoin the front or rear of a house, a quick method can be used to assess the diffuse skylight impact on the house next door. It applies only where the nearest side of the extension is perpendicular to the window (Figure 7); it is not valid for windows which directly face the extension, or for buildings opposite. For these cases the guidelines, in the left-hand column of this page, should be used.

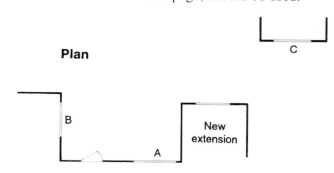

Plan

Figure 7 To assess the impact of the new extension, the 45° approach may be used for window A but not for windows B and C which directly face it

Figure 8 illustrates the application of the method, the '45° approach'. Take the elevation of the window wall and draw diagonally down at an angle of 45° away from the near top corner of the extension. Then take the plan and draw diagonally back at an angle of 45° towards the window wall from the end of the extension. (Note that the section perpendicular to the window is not used here.) If the centre of a main window of the next-door property lies on the extension side of both these 45° lines then the extension may well cause a significant reduction in the skylight received by the window. (In the case of a floor-to-ceiling window, such as a patio door, a point 2 m above the ground on the centre line of the window may be used.)

Like most rules of thumb, this one needs to be interpreted flexibly. For example, if the extension has a much larger building behind it then the daylight from that direction may be blocked anyway. If the extension has a pitched roof then the top of the extension can be taken as the height of its roof halfway along the slope (Figure 8). Special care needs to be taken in cases where an extension already exists on the other side of the window, to avoid a tunnel

effect (Figure 9); it is then advisable to plot the no-sky line in the obstructed room (as already described). Finally, as with the other guidelines in this Section, the 45° approach deals with diffuse skylight only. Additional checks will need to be made for the sunlight which may be blocked.

The windows of some existing buildings will also have rights to light. None of the guidelines here is intended to replace, or be a means of satisfying, the legal requirements contained in rights-to-light law.

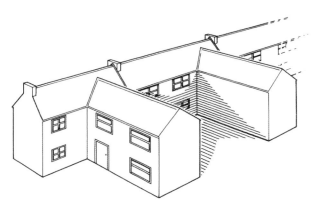

Figure 9 A tunnel effect can occur if windows are obstructed by extensions on both sides

The criterion used in rights-to-light cases is very much a minimum standard, so it is usually true that if the guidelines given here are satisfied then a new development will not infringe rights to light. But this is not always true, particularly if the existing building is unusually deep or has especially small or low windows. If an existing building does have rights to light, and this will usually be the case if it is more than 20 years old, then the designer of the new development should check that it does not infringe them. Appendix E gives further details.

Obstruction of light from the sky is just one of the ways in which a new development can affect existing buildings nearby. The obstruction of sunlight is also important (see Sections 3.2 and 3.3) as are questions of view and privacy (see Section 5).

Elevation

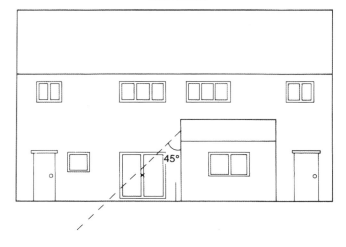

Plan

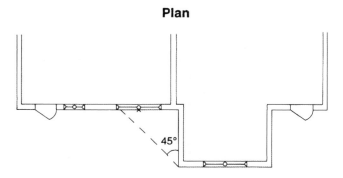

Figure 8 Application of the 45° approach to a domestic extension. A significant amount of light is likely to be blocked if the centre of the window (or, for a floor-to-ceiling window as here, a point 2 m from the ground) lies within the 45° lines on both plan and elevation

Summary (see Figure 10)
If any part of a new building or extension, measured in a vertical section perpendicular to a main window wall of an existing building, from the centre of the lowest window, subtends an angle of more than 25° to the horizontal, then the diffuse daylighting of the existing building may be adversely affected. This will be the case if either:

● the vertical sky component measured at the centre of an existing main window is less than 27%, and less than 0.8 times its former value;

or

● the area of the working plane in a room which can receive direct skylight is reduced to less than 0.8 times its former value.

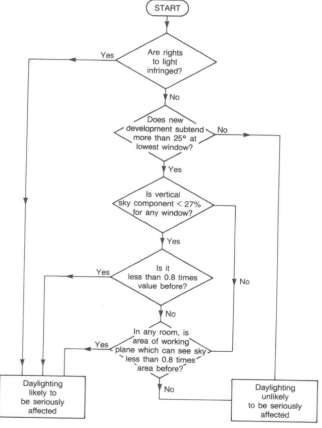

Figure 10 Decision chart; diffuse daylight in existing buildings

2.3 Adjoining development land

From a daylighting standpoint it is possible to reduce the quality of adjoining land by building too close to the boundary. A well designed building will stand a reasonable distance back from the boundaries so as to enable future nearby developments to enjoy similar access to daylight. By doing so it will also keep its own natural light when the adjoining land is developed.

This applies to future non-domestic development as well as housing. However, it does not apply when no main window wall, either of the current new development or of any probable future development on the adjoining site, will face over the boundary. The guidance does not, therefore, apply to a boundary next to a windowless flank wall of a new house where any future housing next door should also present a flank wall without windows; nor need it apply to an industrial estate where new development and any future development is either windowless or solely rooflit.

The diffuse daylight coming over the boundary may be quantified in the following way. As a first check, draw a section in a plane perpendicular to the boundary (Figure 11). If a road separates the two sites, then the centre line of the road should be taken. Measure the angle to the horizontal subtended at a point 2 m above the boundary by the proposed new buildings. If this angle is less than 43° then there will normally still be the potential for good daylighting on the adjoining development site.

If any of the new buildings is taller than this, enough

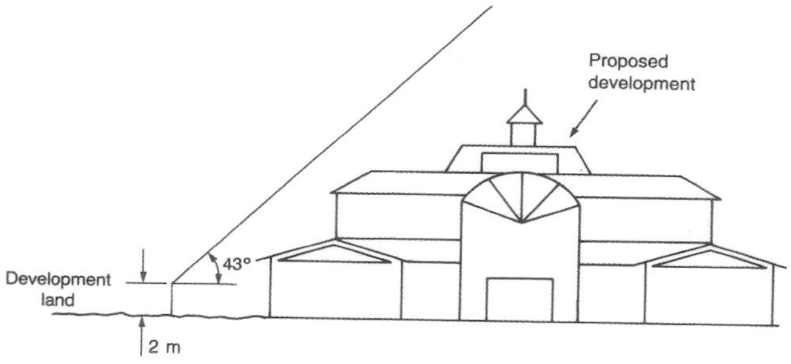

Figure 11 Angular criterion for overshadowing of future development land (on left)

skylight may still reach the development site provided the building is narrow enough to allow adequate light around its sides. This may be quantified by calculating the vertical sky component (see Section 2.1) at a series of points 2 m above the boundary and facing towards the proposed new buildings. Only obstructions caused by the proposed new buildings need to be taken into account. This contrasts with the calculations for buildings where all obstructions need to be included in the analysis. Vertical sky components may be found using the skylight indicator (Appendix A) or Waldram Diagram (Appendix B). Overall, the adjoining development site should normally retain the potential for good daylighting if every point 2 m above the boundary line is within 4 m

7

(measured along the boundary) of a point with a vertical sky component of 17% or more. This corresponds to the value for a continuous obstruction subtending the 43° angle already mentioned.

These guidelines should not be applied too rigidly. A particularly important exception occurs when the two sites are very unequal in size and the proposed new building is larger in scale than the likely future development nearby. The numerical values given in the guidelines are derived by assuming the future development will be exactly the same size as the proposed new building (Figure 12). If the adjoining sites for development are a lot smaller, a better approach is to make a rough prediction of where the nearest window wall of the future development may be; then to carry out the 'new building' analysis in Section 2.1 for this window wall.

The 43° angle should not be used as a form generator,

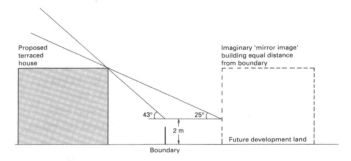

Figure 12 Derivation of an angular boundary criterion to safeguard future development of adjoining land

to produce a building which slopes or steps down towards the boundary. Compare Figure 13 with Figure 12 to see how this can result in a higher than anticipated obstruction to daylight. In Figure 13 the proposed building subtends 34° at its mirror image, rather than the maximum of 25° suggested here. In cases of doubt, the best approach is again to carry out a 'new building' analysis for the most likely location of a window wall of a future development.

The numerical values quoted here are purely advisory. Different values may be used, depending on the type of development earmarked for the adjoining land. All the calculation methods are flexible in this

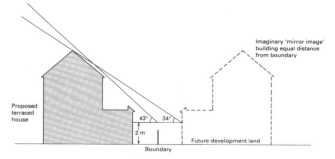

Figure 13 Problems with the boundary criterion can occur when a stepped facade overlooks adjoining land

respect. Table F1 in Appendix F gives the vertical sky components which correspond to different obstruction angles at the boundary, and relates the boundary values to those for faces of buildings to ensure self-consistency.

For simplicity, no numerical guidance is given on sunlighting of land for future development. However a proposed building or group of buildings can significantly reduce the sunlighting of an adjoining site. If this is likely to be a problem, a good way to assess it is to draw the shadows cast by the new buildings at different times of year. Section 3 gives details.

Summary
In broad terms (taking into account the exceptions already noted), a development site next to a proposed new building will retain the potential for good diffuse daylighting provided that on each common boundary:

(a) no new building, measured in a vertical section perpendicular to the boundary, from a point 2 m above ground level, subtends an angle of more than 43° to the horizontal;

or

(b) if (a) is not satisfied, then all points 2 m above the boundary line are within 4 m (measured along the boundary) of a point which has a vertical sky component, looking towards the new building(s), of 17% or more.

3 Sunlighting

Photo courtesy of English Heritage

Figure 14 Stonehenge: an early example of site layout planning for sunlight

3.1 New development

People like sunlight. In surveys, about 90% of those asked said they appreciated having sunlight in their homes. The sun is seen as providing light and warmth, making rooms look bright and cheerful and also having a therapeutic, health giving effect.

In housing, the main requirement for sunlight is in living rooms, where it is valued at any time of day but especially in the afternoon. Sunlight is also required in conservatories. It is viewed as less important in bedrooms and in kitchens, where people prefer it in the morning rather than the afternoon.

Sunlight is also valued in non-domestic buildings. A survey of office workers suggested that 73% of them considered sunlight a pleasure rather than a nuisance. However, the requirement for sunlight will vary according to the type of non-domestic building, the aims of the designer and the extent to which the occupants can control their environment. People appreciate sunlight more if they can choose whether or not to be exposed to it, either by changing their positions in the room or by using adjustable shading. Where prolonged access to sunlight is available, shading devices will also be needed to avoid overheating and unwanted glare from the sun. This can apply to housing as well (Figure 15).

Figure 15 In this housing association scheme, solar shading is provided by balconies and overhangs

In the winter heating season, solar heat gain can be a valuable resource, reducing the need for space heating. Good design can make the most of this. This aspect of sunlight provision is dealt with in Section 4; here we concentrate on the amenity aspects of sunlight.

Site layout is the most important factor affecting the duration of sunlight in buildings. It can be divided into two main issues, orientation and overshadowing.

Orientation

A south-facing window will, in general, receive most sunlight, while a north-facing one will receive it on only a handful of occasions (early morning and late evening in summer). East- and west-facing windows will receive sunlight only at certain times of the day. A dwelling with no main window wall within 90° of due south is likely to be perceived as insufficiently sunlit. This is usually an issue only for flats. Sensitive layout design of flats will ensure that each dwelling has at least one main living room which can receive a reasonable amount of sunlight. In flats and houses a sensible approach is to try to match internal room layout with window wall orientation. Where possible, living rooms should face the southern or western parts of the sky and kitchens towards the north or east.

Overshadowing

The overall access to sunlight of a new development can be considerably enhanced if the layout of new buildings is designed with care so that they overshadow each other as little as possible (see Figure 22 in Section 4). At a simple level, access to sunlight can be improved by:

● Choosing a site on a south-facing slope if possible, rather than a north-facing one

● Having taller buildings to the north of the site with low-rise buildings to the south, but care must be taken not to overshadow neighbouring property (Section 3.2)

● Having low-density housing (semi-detached and detached) at the southern end of a site, with terraced housing to the north

● Placing terraces on east–west roads (so that one window wall faces nearly south) with detached and semi-detached houses on north–south roads

● Opening courtyards to the southern half of the sky

● Having garages to the north of houses

● Avoiding obstructions to the south, such as protruding extensions or other buildings, where window walls face predominantly east or west

● Having low-pitched roofs on housing

For interiors, access to sunlight can be quantified. The British Standard[1] recommends that interiors where the occupants expect sunlight should receive at least one quarter of annual probable sunlight hours, including at least 5% of annual probable sunlight hours during the winter months, between 21 September and 21 March. Here 'probable sunlight hours' means the total number of hours in the year that the sun is expected to shine on unobstructed ground, allowing for average levels of cloudiness for the location in question. The sunlight availability indicator in Appendix A can be used to calculate hours of sunlight received.

At the site layout stage the positions of windows may not have been decided. It is suggested that sunlight availability be checked at points 2 m above the lowest storey level (Figure 1) on each main window wall which faces within 90° of due south. The building face as a whole should have good sunlighting potential if every point on the 2 m high reference line is within 4 m (measured sideways) of a point which meets the British Standard criterion already mentioned[1] for probable sunlight hours. If the access to sunlight changes rapidly along a facade, it is worthwhile trying to site main windows, particularly of living rooms, where most sunlight is available.

If window positions are already known, the centre of each main living room window can be used for the calculation. In the case of a floor-to-ceiling window, a point 2 m above ground on the centre line of the window may be used.

It is not always necessary to do a full calculation to check sunlight potential. It can be shown that the British Standard[1] criterion is met provided either of the following is true:

● The window wall faces within 90° of due south and no obstruction, measured in the section perpendicular to the window wall, subtends an angle of more than 25° to the horizontal (Figure 2 in Section 2.1). Obstructions within 90° of due north of the reference point need not count here.

● The window wall faces within 20° of due south and the reference point has a vertical sky component (Section 2.1) of 27% or more.

The British Standard[1] is intended to give good access to sunlight for amenity purposes in a range of situations. However, in some circumstances the designer or planning authority may wish to choose a different target value for hours of sunlight. This is especially relevant for passive solar buildings, for which Section 4 gives guidance. If sunlight is particularly important in a building, for whatever reason, a higher target value may be chosen, although care needs to be taken to avoid overheating. Conversely, if in a particular development sunlight is deemed to be less important but still worth checking for, a lower target value could be used. In either case, the sunlight availability indicator in Appendix A will show whether the hours of sunlight received meet the target.

Summary

In general, a dwelling or non-domestic building which has a particular requirement for sunlight, will appear reasonably sunlit provided that:

● at least one main window wall faces within 90° of due south;

and

● on this window wall, all points on a line 2 m above ground level are within 4 m (measured sideways) of a point which receives at least a quarter of annual probable sunlight hours, including at least 5% of annual probable sunlight hours during the winter months, between 21 September and 21 March.

3.2 Existing buildings

In designing a new development or extension to a building, take care to safeguard the access to sunlight, both for existing dwellings, and for any nearby non-domestic buildings where there is a particular requirement for sunlight. People are particularly likely to notice a loss of sunlight to their homes, and if it is extensive then it will usually be resented.

Obstruction to sunlight may become an issue if:

● Some part of a new development is situated within 90° of due south of a main window wall of an existing building (Figure 16);

● In the section drawn perpendicular to this existing window wall, the new development subtends an angle greater than 25° to the horizontal measured from a point 2 m above the ground (Figure 2).

To find out whether an existing building still receives enough sunlight, the British Standard[1] can be used. It is suggested that all main living rooms of dwellings, and conservatories, should be checked if they have a window facing within 90° of due south. Kitchens and bedrooms are less important, although care should be taken not to block too much sun. In non-domestic buildings any spaces which are deemed to have a special requirement for sunlight should be checked; they will normally face within 90° of due south anyway.

Access to sunlight should be checked for the main window of each room which faces within 90° of due south. The British Standard[1] recommends that a 'window reference point', at the centre of each

window on the plane of the inside surface of the wall, should be used for the calculations. Sunlight which would be blocked by the window reveals does not count. In the case of a floor-to-ceiling window, such as a patio door, a point on the centre line of the window 2 m above the ground may be used (again on the plane of the inside surface of the wall).

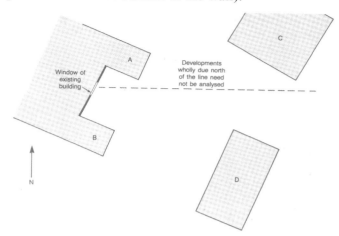

Figure 16 In analysing the sunlighting impact on the existing window, no check need be made for proposed extension A and new building C, as they lie within 90° of due north of the window. Proposed extension B should be checked, as should new building D if it subtends more than 25° to the horizontal, measured in section from the centre of the window

If this window reference point can receive more than one quarter of annual probable sunlight hours (see Section 3.1), including at least 5% of annual probable sunlight hours during the winter months between 21 September and 21 March, then the room should still receive enough sunlight. The sunlight availability indicator in Appendix A, or the rules-of-thumb in Section 3.1, can be used to check this.

Any reduction in sunlight access below this level should be kept to a minimum. If the available sunlight hours are both less than the amount given and less than 0.8 times their former value, either over the whole year or just during the winter months (21 September to 21 March), then the occupants of the existing building will notice the loss of sunlight. The room may appear colder and less cheerful and pleasant.

In certain situations care needs to be taken in applying these guidelines. For example, if the proposed new development is one of a number of successive extensions to the same building, then the total impact on sunlight of all the extensions should be assessed. On the other hand, if the existing building stands unusually close to the common boundary with the new development, then a greater reduction in sunlight access may be unavoidable. The guidelines are purely advisory. Planning authorities may wish to use different criteria, based on the requirements for sunlight in particular types of development in particular areas.

It is good practice to check the sunlighting of gardens of existing buildings. This is described in the next Section.

Summary
If a living room of an existing dwelling has a main window facing within 90° of due south, and any part of a new development subtends an angle of more than 25° to the horizontal measured from the centre of the window in a vertical section perpendicular to the window, then the sunlighting of the existing dwelling may be adversely affected. This will be the case if a point at the centre of the window, in the plane of the inner window wall, receives in the year less than one quarter of annual probable sunlight hours including at least 5% of annual probable sunlight hours between 21 September and 21 March, and less than 0.8 times its former sunlight hours during either period.

3.3 Gardens and open spaces

Good site layout planning for daylight and sunlight should not limit itself to providing good natural lighting inside buildings. Sunlight in the spaces between buildings has an important impact on the overall appearance and ambience of a development.

It is valuable for a number of reasons:

- To provide attractive sunlit views (all year)
- To make outdoor activities like sitting out and children's play more pleasant (mainly during the warmer months)
- To encourage plant growth (mainly in spring and summer)
- To dry out the ground, reducing moss and slime (mainly during the colder months)
- To melt frost, ice and snow (in winter)
- To dry clothes (all year)

The sunlit nature of a site can be enhanced by using some of the techniques described in the previous Section. This could include siting low-rise, low-density housing to the south, with taller, higher density housing to the north of a site; and by opening out courtyards to the southern half of the sky. Special care needs to be taken in the design of courtyards, otherwise they can turn out to be sunless and unappealing (Figure 17).

The use of specific parts of a site can be planned with sunlight in mind. This could include reserving the sunniest parts of the site for gardens and sitting out, while using the shadier areas for car parking. In summer, is often valued in car parks (Figure 18).

Figure 17 Extensive shadowing can occur in courtyards unless care is taken in their design

Figure 18 Shadier areas can usefully be reserved for car parking

The availability of sunlight should be checked for all open spaces where it will be required. This would normally include:

- Gardens, usually the main back garden of a house, and allotments
- Parks and playing fields
- Children's playgrounds
- Outdoor swimming pools and paddling pools
- Sitting-out areas, such as those between non-domestic buildings and in public squares

● Focal points for views, such as a group of monuments or fountains

Each of these spaces will have different sunlighting requirements and it is difficult to suggest a hard and fast rule. However, it is clear that the worst situation is to have significant areas on which the sun does not shine for a large part of the year. These areas will, in general, be damp, chilly and uninviting (Figure 19). The equinox (21 March) is a good date for assessment.

This problem occurs with only certain forms of layout. If a long face of a building faces within 13° of due north, then there will be an area adjoining the building face which is permanently in shade at the equinox (and hence all winter). Areas of this sort can also occur if buildings form an enclosed or partly enclosed space which is blocked off from the southern half of the sky. Figure 20 illustrates some typical examples.

Figure 19   This outdoor space is in shade all winter. It is grim and underused

It is usually possible to redesign the layout to minimise these areas, either by reorienting buildings or by opening gaps to the south in courtyards.

Where this is not possible, it is suggested that no more than two-fifths, and preferably no more than a quarter, of any of the listed amenity areas should be prevented by buildings from receiving any sunlight at all on 21 March. Sunlight at an altitude of 10° or less does not count. In working out the total area to be considered, driveways and hard standing for cars should be left out. Around housing, front gardens which are relatively small and visible from public footpaths should be omitted; only the main back garden should be analysed. Each individual garden for each dwelling in a block should be considered separately.

Areas of open space which can and cannot receive sunlight on 21 March may be found using the sunlight-on-ground indicator (Appendix G). It is

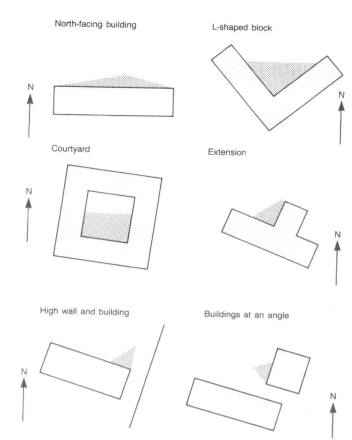

Figure 20   Examples of layouts where poor sunlighting on the ground can occur. The shaded areas will receive no sunlight at the equinox

instructive to draw the no-sun contour which separates these areas on plan. For conventional buildings, if a point lies within the no-sun contour (ie it receives no sun on 21 March), it will be permanently in shade all winter. Likewise, if it can receive some sun on 21 March, it will receive some sunlight all summer. Here 'conventional buildings' means structures without overhangs, open arches or overhead walkways.

The question of whether trees or fences should be included in the calculation depends upon the type of shade they produce. Normally, trees and shrubs need not be included, partly because their shapes are almost impossible to predict, and partly because the dappled shade of a tree is more pleasant than the deep shadow of a building. This applies especially to deciduous trees. Nevertheless, choose locations for tree planting with care. The aim should normally be to have some areas of partial shade under trees while leaving other parts of the garden or amenity area in full sun. Where a dense belt or group of evergreens is specifically planned as a windbreak or for privacy purposes, it is better to include them in the calculation of shaded area (Figure 21). The growth of trees and their likely final size should be allowed for.

Fences and walls cast deeper shade than trees, and their positions can often be predicted. As a guide, it is suggested that where the plan calls for walls or opaque fences greater than 1.5 m high, the shadows they cast should be included in the calculation. Where low fences or walls are intended, or no specific provision is made, no calculation of shadows is necessary.

Figure 21 A dense belt of coniferous trees should be treated as an obstruction to sunlight

This guidance applies both to new gardens and amenity areas and to existing ones which are affected by new developments. If an existing garden or outdoor space is already heavily obstructed, then any further loss of sunlight should be kept to a minimum. In this poorly sunlit case, if as a result of new development the area which can receive direct sunlight on 21 March is reduced to less than 0.8 times its former size, then this further loss of sunlight is significant. The garden or amenity area will tend to look more heavily overshadowed.

It is important to realise that the area-based guideline is very much a minimum standard. It will not guarantee large amounts of sun in summer, or any sun at all in winter. It will not ensure that sunlight is available in specific areas like patios, terraces or flower beds. For critical areas it is suggested that a more detailed study of sunlighting potential be carried out, using a prediction tool such as the sunpath indicator in Appendix A, or BRE's *Sunlight availability protractor* (see back cover).

It is also important to use the guideline sensibly. There is little point in leaving a tiny gap between

buildings so that a thin shaft of sunlight penetrates through to a gloomy 'amenity area' on 21 March.

Where a large building is proposed which may affect a number of gardens or open spaces, it is often illustrative to plot a shadow plan showing the location of shadows at different times of day and year. For 21 March this can be done by using the sun-on-ground indicator in reverse (Appendix G).

Summary

It is suggested that, for it to appear adequately sunlit throughout the year, no more than two-fifths and preferably no more than a quarter of any garden or amenity area should be prevented by buildings from receiving any sun at all on 21 March. If, as a result of new development, an existing garden or amenity area does not meet these guidelines, and the area which can receive some sun on 21 March is less than 0.8 times its former value, then the loss of sunlight is likely to be noticeable.

4 Passive solar design

As well as bringing warmth and vitality to exterior and interior spaces, the sun is also a source of energy. Good building design should seek to tap this energy to reduce consumption of conventional fuels. Where this becomes a particular priority in arranging the form, fabric and systems of a building and the site layout, the result is a passive solar design. Passive solar homes can have a heating energy consumption up to 2000 kWh a year lower than conventional housing, depending on their size (ETSU figures[3]). These benefits depend upon the arrangement of the site to produce the best orientation (closest to the south) and to reduce overshadowing (Figure 22).

Even houses with no special design features benefit from solar energy (up to 500 kWh/year) if oriented in a north–south direction without overshadowing.

This Section will concentrate on passive solar design rather than active systems like solar panels. This is partly because, in the United Kingdom at least, passive design tends to be more cost-effective, and partly because solar gain to roof-mounted panels is less influenced by site layout as it is less likely to be blocked by other buildings. However, much of the advice given here is valid also for active systems, except that any overshadowing checks will need to be done at solar panel level rather than window level.

Passive solar design is valuable for non-domestic buildings as well as for housing. In the non-domestic sector, Duncan and Hawkes[4] have estimated that passive solar energy could displace over 1 million tonnes of coal equivalent per year in the UK. Around half of this would be the result of savings in lighting energy consumption. (A BRE Report reaches similar conclusions[5].) The guidance which follows could be applied also to non-domestic buildings, although greater care needs to be taken to avoid overheating. However, the provision of adequate daylight and electric lighting controls is especially important (refer to Section 2.1).

In deciding whether to opt for passive solar design, the needs of the client and the intended use of the building are important. Site-related factors will also influence the decision. On a sloping site which faces north, it will be harder to reap the full benefits of passive solar design; conversely, a south-facing slope will make it easier. At unusually high densities of development it becomes difficult to avoid serious obstruction or poor orientation for at least some of the houses. Similarly, on a small site it may be impossible to achieve the best orientation for window walls or to avoid overshadowing by nearby buildings.

These factors need to be carefully considered in passive solar design if the potential energy savings are to be realised. An alternative approach is to concentrate on providing daylight and sunlight as an amenity (see Sections 2 and 3), and perhaps to introduce other energy measures such as improved insulation. The following guidance is intended for those buildings which are specifically designed to make the most of ambient solar energy, when it is intended to supplement the advice in Sections 2 and 3. Passive solar site layout design can be divided into the two key issues: orientation and overshadowing.

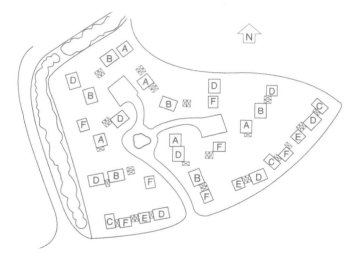

Figure 22 A site layout design study by NBA Tectonics for ETSU[3]. The conventional layout of detached houses (top) would need 8900 kWh/year for space heating, 8500 kWh/year with passive solar features. The passive solar site layout (bottom), redesigned by Stillman Eastwick-Field, would require only 7900 kWh/year, a saving of over 10%

Orientation
To make the most of solar gain, the main solar collecting facades of domestic buildings should face within 30° of due south. Orientations further east or west than this will receive less solar gain, particularly in winter when it is of most use.

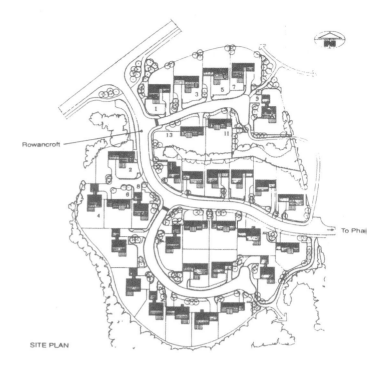

Rowancroft

To Pha

SITE PLAN

Figure 23 At Willow Park, Chorley (top), careful road layout design (right) means that all the passive solar homes can have a southerly orientation

These orientation requirements have considerable influence on site layout. A variety of design solutions is possible, but careful design is needed to offset the monotony that could result from a majority of houses facing south. To achieve a variety of form and spaces, traditional strategies can be used, such as mixing house types, varying the siting within house plots, and good landscaping. Roads will ideally be east–west, but other solutions are possible (Figure 23).

The individual layout of each building will also be affected. In houses, the solar gain will be used most effectively if living rooms are sited on the south side, with kitchens, bathrooms and garages to the north (Figure 24). In non-domestic buildings, toilets, storerooms, computer rooms, canteens and other rooms with high internal gains can be located to the north.

Overshadowing

Care must be taken to ensure that overshadowing by other buildings does not lessen the effectiveness of a passive solar design. A solar collecting facade needs access to low-angle sun in winter when its contribution will be most valuable (Figures 25(a) and 25(b)).

Photo by courtesy of CEC Project Monitor[6]

Figure 25(a) Passive solar homes at Giffard Park, Milton Keynes

Figure 24 Ground-floor plan of Linford low-energy house, Milton Keynes

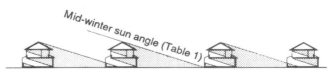

Figure 25(b) The terraces at Giffard Park are carefully spaced to avoid winter overshadowing

Overshadowing can be minimised by adopting the measures listed in Section 3.1. These include having taller buildings and high-density development (such as terraces) to the north of the site, with lower-rise, low-density development (such as bungalows and detached houses) to the south. Terraces can be placed on east–west roads so that one window wall faces south. Where necessary, detached houses can be located on north–south roads. Roof slopes can be reduced to increase solar access to buildings to the north.

It is also possible to choose plot shapes and the locations of buildings within them to minimise overshading. Tree locations are important; deciduous species are best because they are leafless when solar gains are most valuable, while providing some shade in summer.

To reap the full benefits of passive solar design,

maximise winter solar gain as far as other site layout constraints allow. For this purpose the most important area to keep free of obstruction is within 30° of due south of a solar collecting facade (Figure 26). This is the part of the sky from which most solar radiation comes in the winter months. To check whether solar access from this zone is retained, draw a north–south section (not necessarily perpendicular to the facade). The altitude of any obstructions in it should not exceed the critical angle h when measured from the centre of the solar collecting glazing. Values of h are given in Table 1. If the obstruction angle does not exceed h, then at least three hours of sunlight around midday are guaranteed for the period specified — provided the sun shines, of course. Note that the values of h are given in terms of site latitude, so if solar gain were required all year at a site in Cardiff (51.5° N) then the maximum obstruction angle h in Figure 26 would be 65° – 51.5° = 13.5°.

Section O – S

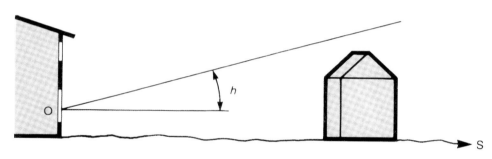

Plan

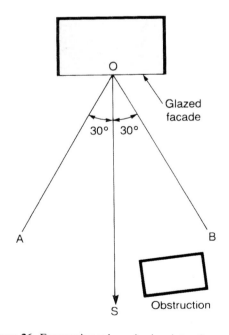

Table 1 Limiting obstruction angles *h* to ensure at least three hours' possible sun per day in specified period

Period of year	London	Manchester	Edinburgh	Other location
		Value of *h*		
All year	13°	11°	9°	65° minus latitude
21 Jan –21 Nov	17°	15°	12°	68° minus latitude
6 Feb –6 Nov	21°	19°	16°	72° minus latitude
21 Feb –21 Oct	27°	25°	22°	78° minus latitude

Figure 26 For passive solar gains in winter the sector AOB 30° either side of due south is important. To guarantee winter sun from this sector obstructions within it should not subtend more than the critical angle h when measured in section. Table 1 gives values for h

It may not always be possible or desirable to plan for this amount of open space in front of glazed areas, particularly in Scotland, Northern Ireland and the north of England. If an obstruction takes up only part of the ±30° angular zone either side of due south, enough solar gain may still be available from other directions. To check this, the solar gain indicator in Appendix A can be used. This will give the solar radiation, in kWh/m², falling on an obstructed south-facing window in the winter heating season. It is possible to get both an absolute amount of radiation and a comparison with the value for a totally unobstructed window. It is intended only for vertical windows facing approximately due south, and should not be used for sloping glazing or solar panels, or where the window faces more than 30° east or west of due south.

It is also important to check whether a passive solar building receives enough diffuse daylight (Section 2.1). This may affect the energy efficiency of the building as well as its attractiveness to the occupants. Special care should be taken to ensure good daylighting to the north side of the building, as often minimal window areas are chosen on thermal grounds (Figure 27).

Where a proposed development of any type is near to an existing building designed to make use of solar gain, it is good practice to try to minimise any loss of that solar gain. This can be done using the solar gain indicator in Appendix A.

However, when designing a passive solar building the possibility of future development blocking solar access should be anticipated. The building(s) should stand well back from the southern boundary of the site unless it is clear that no future development is to take place there. It is neither reasonable nor prudent to build close to a southern boundary and expect future development to stand well back from it.

When a solar estate is constructed it is wise to guard against the possibility of neighbours blocking each other's solar radiation by erecting large extensions and outbuildings. This can be done by drawing up legal agreements which protect solar access for each property. It is necessary to have an unambiguous numerical criterion for solar access. This could be in terms of the winter solar radiation as calculated using the solar gain indicator in Appendix A.

Alternatively, it could be quantified as the number of possible hours of sunlight on a particular day or days. The sunpath indicator in Appendix A can be used to check this. Whatever criterion is chosen, it should be realistic. It should rule out future buildings which would cause a serious reduction in solar gain to an adjoining property, but not prevent smaller buildings which would have little or no impact on solar access. And, obviously, the criterion should be met with the solar estate as initially constructed; careful checks should be made to ensure this.

Summary
Passive solar buildings are designed to make the most of ambient solar energy. In general, they will have adequate solar access if:

● the main solar collecting glazing faces within 30° of due south;

and

● the sector on plan within ±30° of due south from this glazing is kept only lightly obstructed. To receive sunlight from this zone for specified periods of the year, the obstruction angle measured in a north–south section from the centre of the glazing should not exceed the values given in Table 1. As an alternative, the solar radiation received by the glazing during the winter heating season may be found, using the solar gain indicator in Appendix A.

Photo by courtesy of CEC Project Monitor[6]

Figure 27 In these passive solar homes, a variety of front door positions and a mixture of garages and car ports produce attractive north facades

18

5 Other issues

5.1 Introduction

Daylight and sunlight are only two of the issues that need to be considered at the site layout stage. This Section briefly mentions other issues which may have an impact on the natural lighting of a layout. Further information can be found in the Bibliography.

It should be noted that while the current (1991) Building Regulations do not contain requirements on daylight or sunlight, Part B[7] on fire spread contains some constraints on the minimum spacing of buildings from boundaries. A well spaced layout will enable buildings to be reasonably daylit without unduly large window areas, thus helping to meet the thermal performance criteria in Part L[7].

5.2 View

At the site layout stage in design, the needs for view and for daylight rarely conflict; an open, well daylit layout will usually provide reasonable views. There is often a need for specific, close views near to a building, for example for supervision of children playing in a garden, or for security reasons (see Section 5.4). In this case windows may be needed where they will not necessarily receive the most daylight or sunlight.

This may constrain the design of passive solar homes. Rooms on the north side of such homes should have enough window area to provide reasonable views out, where these are required. This particularly applies to kitchens, which on thermal grounds would normally be sited away from the well glazed south facade.

5.3 Privacy

Privacy of houses and gardens is a major issue in domestic site layout. Overlooking from public roads and paths and from other dwellings needs to be considered. The way in which privacy is achieved will have a major impact on the natural lighting of a layout. One way is by remoteness; by arranging for enough distance between buildings, especially where two sets of windows face each other. Recommended privacy distances in this situation vary widely, typically from 18 m up to 35 m. A spacing-to-height ratio of just over 2:1 is normally enough to allow adequate daylighting on building faces; thus, for low-rise housing, if these privacy distances are applied good natural lighting in the layout will ensue automatically. However, smaller-scale checks, for example, of overshadowing by extensions, may still be necessary.

The second way of achieving privacy is by design. High walls, projecting wings and outbuildings block direct views of interiors, and some windows may be diffusing or at high level. In this situation, natural lighting is often reduced, both because the visual screens themselves block it, and because the spacing between buildings may be much less. To achieve good sunlight and daylight in this type of layout the guidance in Sections 2 and 3 needs to be followed carefully.

Privacy may be an issue in passive solar dwellings. If houses have little north-facing glazing it may be possible to reduce the spacing between them without enabling overlooking between dwellings. In this case it is necessary to check whether the south sides of the houses still receive enough solar gain (Section 4 gives details). Other problems may arise because of the large areas of glazing on the south side of the passive solar home, and consequent loss of privacy for living rooms. Gardens may need to be extended and access roads and paths situated so as to avoid overlooking. In tight layouts, one compromise is to have a conservatory on the south wall. Then, although it may be possible to see into the conservatory, it will be less easy to see into the living rooms behind it.

5.4 Security

There may be occasional conflicts between site layout design to provide sunlight and daylight, and security requirements. These may occur in a number of ways.

- To ensure good overall sunlight and daylight in a housing estate it is usually better to space out dwellings evenly, while grouping homes in small clusters promotes neighbourliness and natural surveillance.

- It may be necessary to erect high walls or fences, for example where the rear of a property faces open ground.

- Windows may need to be positioned so that occupants can view areas immediately adjacent to a building. These may not be the best positions for access to sunlight and daylight.

- For maximum sunlight in gardens and planted areas it is best to park cars in shaded areas. However, for security cars should be easily seen from the occupied building.

These conflicts can usually be resolved by careful site layout design. For housing, a Guidance Note of the National House-Building Council (NHBC)[8] gives advice on site layout for security .

5.5 Access

With careful site layout planning it is possible to

satisfy the needs both for pedestrian and vehicle access, and for adequate sunlight and daylight. The spaces required by roads and footpaths will bring sunlight and daylight into a layout. Tall buildings can be sited to the south of larger road junctions, where they will cast shadows on roads rather than on other buildings.

Nevertheless, there is often a three-way conflict between good natural lighting, access and privacy (Section 5.3). For maximum privacy, large windows should not face roads or other public spaces, even though they would probably receive most sunlight and daylight there. In practice, this problem may be overcome by having private zones such as front gardens, or shading devices like net curtains.

Problems with road and footpath layouts may occur in passive solar estates. The best type of road pattern for solar access is a series of long east–west roads with shorter north–south link roads. However, in residential areas shorter, curving roads are usually favoured because this will reduce traffic speed and produce a succession of smaller, more restful looking spaces. Such a road layout will require greater imagination in the design of passive solar housing (Figure 23). Detached houses or houses with roof collectors may be used on north–south roads, perhaps with gable ends facing the road. A report by NBA Tectonics for ETSU[3] gives some practical ideas.

5.6 Enclosure

To achieve good natural lighting within a site, there needs to be enough space between buildings. However, in built-up areas the perceived quality of an outdoor space may be reduced if it is too wide and long compared to the height of the buildings that surround it. It may lack human scale and a sense of enclosure (Figure 28).

There is clearly a conflict here. One way of resolving it is actually to increase the space between buildings and to use landscaping and planting to disguise them. Then the view appears less as a space surrounded by buildings and more as a natural landscape with the occasional building. However, this is an option only if the density of buildings is very low.

For higher densities it is still possible to retain a sense of enclosure along with reasonable sunlight and daylight. In linear spaces, such as a street between two rows of terraces, a spacing-to-height ratio of 2.5:1 would still appear enclosed, but not obstruct too much natural light. The worst conflicts can occur in courtyards. It has been suggested that to appear as an enclosed space a courtyard should have a spacing-to-height ratio of 4:1 or less. However, courtyards of this shape, completely enclosed on all four sides, will have less than ideal natural lighting. Rooms lit by windows

Figure 28 Outdoor spaces can have little sense of enclosure (top) or a strong sense of enclosure (bottom)

near the corners of the courtyard may appear gloomy and heavily obstructed. Sunlight will, at most, reach only half the courtyard in winter.

Such problems can be overcome in a number of ways:

● Including gaps between buildings, especially on the south side of a courtyard, will improve access to sunlight and daylight. If the gaps are not too large the space will still appear reasonably enclosed.

● A larger space between buildings can be broken up, by vegetation or well defined changes in ground level, into what appears to be a number of smaller spaces.

● Reducing both the height and width of an outdoor space, while keeping the ratio of the two the same, will increase the sense of enclosure while slightly increasing the natural light available at window head level.

● In some circumstances the need for daylight at ground floor level may not be great; for example, where shops occupy the ground floor. Alternatively, the ground floor areas could be reduced and lit primarily from the less obstructed side (Figure 29).

Figure 29 In this enclosed mews, the ground floor area of each house has been reduced. The windows by the front doors are for view and security purposes only

- Smaller spacings could be compensated for by increasing window size, especially window head height, and decreasing room depth. This is the approach adopted in Georgian buildings (Figure 5). Appendices C and F give advice on how to achieve this.

5.7 Microclimate

Access to daylight and sunlight is an important aspect of the microclimate around buildings. The other main element of microclimate is shelter, either from the wind or from excessive solar heat gain in summer. There may be a conflict between shelter and solar access requirements. A BRE Digest[9] deals with this issue in detail; only an outline is given here.

Measures to provide shelter from the wind may compromise access to natural light, for example:

- The planting or construction of windbreaks

- The plugging of gaps between buildings through which wind could rush
- The reduction of distances between buildings so that they partly shield each other

A compromise solution will depend on how exposed the site is. On a particularly exposed site it may be necessary to plant rows of conifers for use as windbreaks, even if this reduces levels of daylight (Figure 30). Figure 31 shows ways to make individual buildings less sensitive to wind without necessarily affecting natural light.

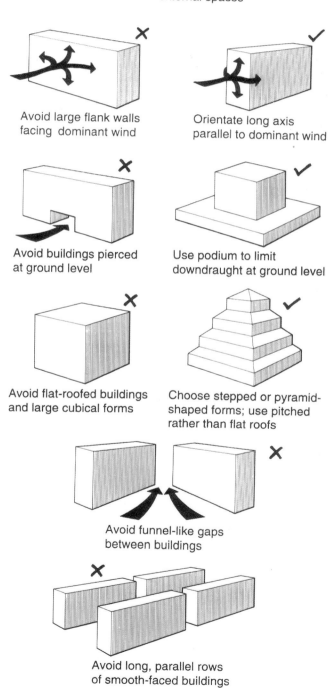

Choose form and arrangement of building to avoid downdraughts and shelter external spaces

Avoid large flank walls facing dominant wind

Orientate long axis parallel to dominant wind

Avoid buildings pierced at ground level

Use podium to limit downdraught at ground level

Avoid flat-roofed buildings and large cubical forms

Choose stepped or pyramid-shaped forms; use pitched rather than flat roofs

Avoid funnel-like gaps between buildings

Avoid long, parallel rows of smooth-faced buildings

Figure 31 Reducing the wind sensitivity of buildings

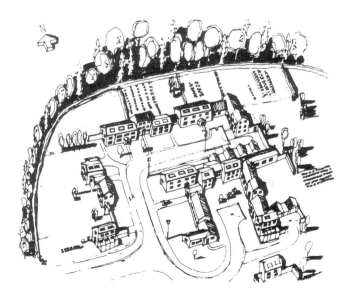

Figure 30 A site design providing substantial tree shelter to north, with good solar access to buildings and spaces

Summertime shade can be provided in a number of ways. Deciduous trees give shade in summer but allow access to sunlight and daylight in winter. Buildings can incorporate shading devices such as overhangs which block high-angle summer sun. Making building surfaces a light colour will reduce absorbed radiation and improve reflected light. Site features like lakes and vegetation can reduce summer temperatures by evaporative cooling and thermal storage.

5.8 Solar dazzle

Glare or dazzle can occur when sunlight is reflected from a glazed facade (Figure 32). This can affect road users outside and the occupants of adjoining buildings. The problem can occur either when there are large areas of reflective tinted glass on the facade, or when there are areas of glass which slope back at up to 35° from the vertical so that high altitude sunlight can be reflected along the ground (Figures 33 and 34). Solar dazzle is a long-term problem only for some heavily glazed (or mirror-clad) non-domestic buildings. A glazed facade also needs to face within 90° of due south for significant amounts of sunlight to be reflected.

If it is likely that a building will cause solar dazzle, the exact scale of the problem should be evaluated. This is done by identifying key locations such as road junctions and windows of nearby buildings, and working out the number of hours of the year that sunlight can be reflected to these points. A BRE Information Paper[10] gives full details.

At the design stage, solar dazzle can be remedied by reducing areas of glazing, substituting clear or absorbing glass for reflective glass, reorienting the building, or replacing areas of tilted glass by either vertical or nearly horizontal glazing. Alternatively some form of opaque screening may be acceptable, although this usually needs to be larger than the glazing area.

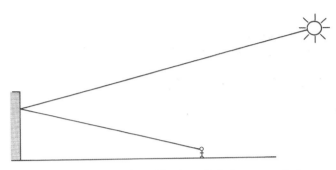

Figure 33 Reflection of low-altitude sunlight from a vertical facade

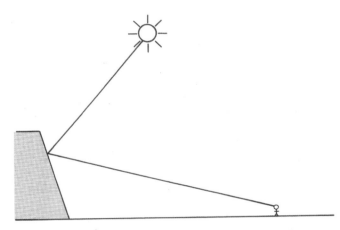

Figure 34 Reflection of high-altitude sunlight from a sloping facade

Figure 32 Solar dazzle reflected from a glazed facade

References

1   **British Standards Institution.** Code of practice for daylighting. *British Standard* BS 8206:Part 2:1992. London, BSI, 1992.

2   **Chartered Institution of Building Services Engineers.** *Applications manual: window design.* London, CIBSE, 1987.

3   **NBA Tectonics.** A study of passive solar housing estate layout. *ETSU Report* S-1126. Harwell, Energy Technology Support Unit (ETSU), 1988.

4   **Duncan I P and Hawkes D.** Passive solar design in non-domestic buildings. *ETSU Report* S–1110. Harwell, Energy Technology Support Unit (ETSU), 1983.

5   **Crisp V H C, Littlefair P J, Cooper I and McKennan G.** *Daylight as a passive solar energy option: an assessment of its potential in non-domestic buildings.* Building Research Establishment Report. Garston, BRE, 1988.

6   **Lewis J O** (Ed). *Project Monitor.* Dublin, Commission of the European Communities/ University College, Dublin, 1987–1989.

7   **Department of the Environment and the Welsh Office.** *The Building Regulations* 1991. London, HMSO, 1991. *Approved Document B: Safety in fire.* 1991. *Approved Documents L: Conservation of fuel and power* — L2/3, L4, L5. 1990.

8   **National House-Building Council.** *Guidance on how the security of new homes can be improved.* Amersham, NHBC, 1986.

9   **Building Research Establishment.** Climate and site development. Part 1: General climate of the UK. Part 2: Influence of microclimate. Part 3: Improving microclimate through design. *BRE Digest* 350. Garston, BRE, 1983.

10 **Littlefair P J.** Solar dazzle reflected from sloping glazed facades. *Building Research Establishment Information Paper* IP3/87. Garston, BRE, 1987.

Bibliography

Daylighting and sunlighting

British Standards Institution. Code of practice for daylighting. *British Standard* BS 8206:Part 2:1992. London, BSI, 1992.

Chartered Institution of Building Services Engineers. *Applications manual: window design.* London, CIBSE, 1987.

Hopkinson R G, Petherbridge P and Longmore J. *Daylighting.* London, Heinemann, 1966.

Lynes J A. *Principles of natural lighting.* London, Elsevier, 1968.

Turner D P (Ed). *Windows and environment.* Newton-le-Willows, McCorquodale/Pilkingtons, 1969.

Moore F. *Concepts and practice of architectural daylighting.* New York, Van Nostrand, 1985.

Crisp V H C, Littlefair P J, Cooper I and McKennan G. *Daylight as a passive solar energy option: an assessment of its potential in non-domestic buildings.* Building Research Establishment Report. Garston, BRE, 1988.

Ne'eman E and Light W. *Availability of sunshine.* Building Research Establishment Current Paper CP75/75. Garston, BRE, 1975.

Building Research Establishment. *Sunlight availability protractor.* Garston, BRE, 1975.

Hunt D R G. *Availability of daylight.* Building Research Establishment Report. Garston, BRE, 1979.

Rights to light

Anstey J. *Rights of light and how to deal with them.* Surveyors Publications. London, Royal Institute of Chartered Surveyors (RICS), 1988.

Ellis P. *Rights to Light.* London, Estates Gazette, 1989.

Passive solar design

NBA Tectonics. A study of passive solar housing estate layout. *ETSU Report* S-1126. Harwell, Energy Technology Support Unit (ETSU), 1988.

Brown G Z. *Sun, wind and light: architectural design strategies.* New York, Wiley, 1985.

Knowles R L. *Sun rhythm form.* Cambridge, Massachusetts, United States, Massachusetts Institute of Technology (MIT) Press, 1981.

Turrent D, Doggart J and Ferraro R. *Passive solar housing in the UK.* London, Energy Conscious Design, 1981.

Lebens R. *Passive solar heating design.* London, Applied Science, 1980.

Lewis J O (Ed). *Project Monitor.* Dublin, Commission of the European Communities/University College, Dublin, 1987–1989.

Robinette G O (Ed). *Energy efficient site design.* New York, Van Nostrand, 1983.

Other issues

Essex County Council. *A design guide for residential areas.* Chelmsford, Essex County Council, 1973.

Greater London Council. *An introduction to housing layout.* London, Architectural Press, 1978.

Department of the Environment. *Housing. Planning Policy Guidance Note* 3. London, 1992.

Woodford G T, Williams K and Hill N. The value of standards for the external residential environment. *Research Report* 6. London, Department of the Environment (DOE), 1976.

Clouston B and Stansfield K (Ed). *Trees in towns.* London, Architectural Press, 1981.

Gruffydd B. *Tree form, size and colour — a guide to selection, planting and design.* London, Spon, 1987.

Property Services Agency. *Energy saving through landscape planning.* Croydon, PSA, 1988. (Available from BRE Bookshop.)

Lisney A, Fieldhouse K and Dodd J. *Landscape design guide (Vols 1, 2).* Aldershot, Property Services Agency (PSA)/Gower, 1990.

National House-Building Council. *Guidance on how the security of new homes can be improved.* Amersham, NHBC, 1986

Department of the Environment. Residential roads and footpaths: layout considerations. *DOE Design Bulletin* 32. London, HMSO, 1977 (under revision).
Cheshire County Council. *Design aid — housing:*

roads, incorporating residential road standards.
Chester, Cheshire County Council, 1976.

Bentley I, Alcock A, Murrain P, McGlynn S and Smith G. *Responsive environments: a manual for designers.* London, Architectural Press, 1985.

Simpson B J and Purdy M T. *Housing on sloping sites: a design guide.* London, Construction Press, 1984.

BRE Digests
Building Research Establishment. Lighting controls and daylight use. *BRE Digest* 272. Garston, BRE, 1983.

Building Research Establishment. Estimating daylight in buildings: Parts 1 and 2. *BRE Digests* 309 and 310. Garston, BRE, 1986.

Building Research Establishment. Climate and site development. Part 1: General climate of the UK. Part 2: Influence of microclimate. Part 3: Improving microclimate through design. *BRE Digest* 350. Garston, BRE, 1983.

BRE Information Papers
Littlefair P J. Solar dazzle reflected from sloping glazed facades. *Building Research Establishment Information Paper* IP3/87. Garston, BRE, 1987.

Slater A I. Lighting controls: an essential element of energy efficiency. *Building Research Establishment Information Paper* IP5/87. Garston, BRE, 1987.

Littlefair P J. Average daylight factor: a simple basis for daylight design. *Building Research Establishment Information Paper* IP15/88. Garston, BRE, 1988.

Littlefair P J. Innovative daylighting systems. *Building Research Establishment Information Paper* IP22/89. Garston, BRE, 1989.

Acknowledgements

This report was produced after an extensive period of consultation with architects, planning officers, consultants, professional institutions, and officials of the Department of the Environment. The contributions of all concerned are gratefully acknowledged. Special thanks must go to Mr J Lynes, of Humberside Polytechnic, and to Mr J Basing, who helped to formulate some of the detailed guidance here. Photographs are reprinted with permission from the Projects Monitor Series produced for the Commission of the European Communities by The ECD Partnership (Figures 23, 25(a) and 27), and from English Heritage (Figure 14).

Appendix A

Indicators to calculate access to skylight, sunlight and solar radiation

A1 General

This Appendix contains indicators to find how much skylight, sunlight and solar gain reach the outside of a window. These comprise:

- The skylight indicator (Figure A1), to find the vertical sky component (in %) on the outside of a window wall (Section 2)

- The sunlight availability indicators (Figures A2 to A4) to find the probable sunlight hours received by a window wall, or at any other point in a building layout (Sections 3.1 and 3.2)

- The sunpath indicators (Figures A5 to A7) to find the times of day and year for which sunlight is available on a window wall or point in a layout

- The solar gain indicators (Figures A8 to A10) to find the incident solar radiation on a south-facing vertical window wall (Section 4)

The last three indicators (sunlight availability, sunpath and solar gain) come in three different versions of each, according to the latitude of the site in question. The indicators marked 'London 51.5° N' may be used for southern England and south Wales. The 'Manchester 53.5° N' ones are for northern England, north Wales and the southern half of Northern Ireland. For Scotland and the northern half of Northern Ireland the 'Edinburgh/Glasgow 56° N' indicators may be used.

The skylight indicator is independent of latitude and may be used anywhere.

The skylight and solar gain indicators are semi-circular; the sunlight availability and sunpath indicators are shaped like a circle with a segment removed. In each case the centre of the circular arc corresponds to the reference point at which the calculation is carried out. Radial distances from this point correspond to the ratio of the distance of the obstruction on plan divided by its height above the reference point. So if the reference point is 2 m above ground, and the ground is flat, this height will be the obstruction height above ground, minus 2 m. The indicators are all drawn to the same scale so that it is easy to calculate a number of different quantities at the same time.

Directions on the indicator from the central point correspond to directions on the site plan. The skylight indicator is used with its straight base parallel to the window wall. The sunlight availability, sunpath, and solar gain indicators, however, are always used with the south point of the indicator pointing in the south direction on plan, whatever the orientation of the window wall.

Unlike previous daylight and sunlight indicators, and the sun-on-ground indicators in Appendix G, these indicators are not transparent. They are not intended to be laid over standard scale site plans, because the distance scale on the indicator is unlikely to correspond to the scale of the plan. To plot a layout on the indicator either the transparent direction finder may be used, or a plan may be specially drawn to the exact scale of the indicator.

Use of the transparent direction finder

This is loosely inserted into the book and is illustrated in Figure A11. The direction finder is placed on the site plan (of whatever scale). Its centre should be at the reference point and its base parallel to the wall. The direction finder is divided into eight radial zones, each of which can be further divided into two if necessary.

Take each radial zone in turn and determine whether there are any obstructions in it. If an obstruction covers only half a particular zone, each half of the zone may be considered separately. Take the average distance of the nearest face of each obstruction in that zone or half zone. For each obstruction, calculate the following ratio — (Distance of obstruction):(Height above reference point). In each zone, if one obstruction lies behind another, choose the obstruction with the lowest value of this ratio. This will be the obstruction that, seen from the reference point, appears to block more of the sky. For that obstruction, mark, on the radial scale of the direction finder, the zone or half zone at the ratio calculated. A washable overhead projector pen can be used to make the marks. It is important to realise that the position of the obstruction on the plan will not usually coincide with its position on the scale of the direction indicator, although its direction will.

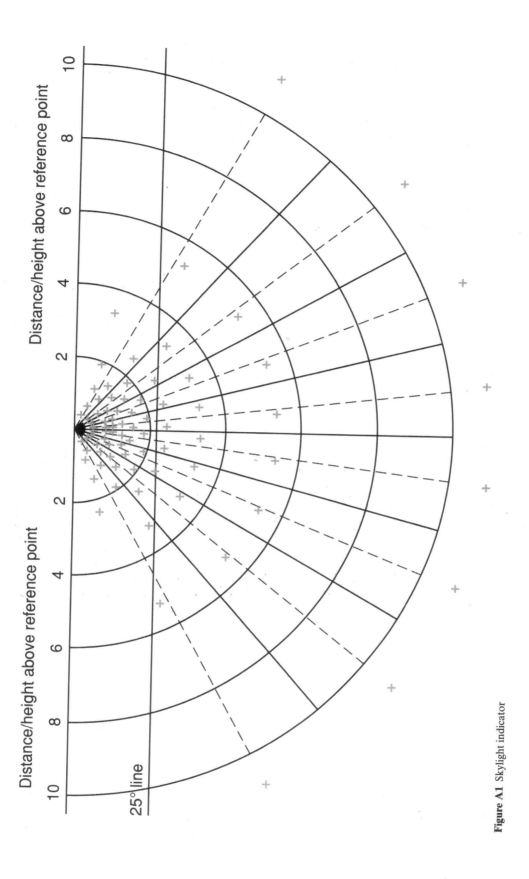

Distance/height above reference point

Distance/height above reference point

25° line

Figure A1 Skylight indicator

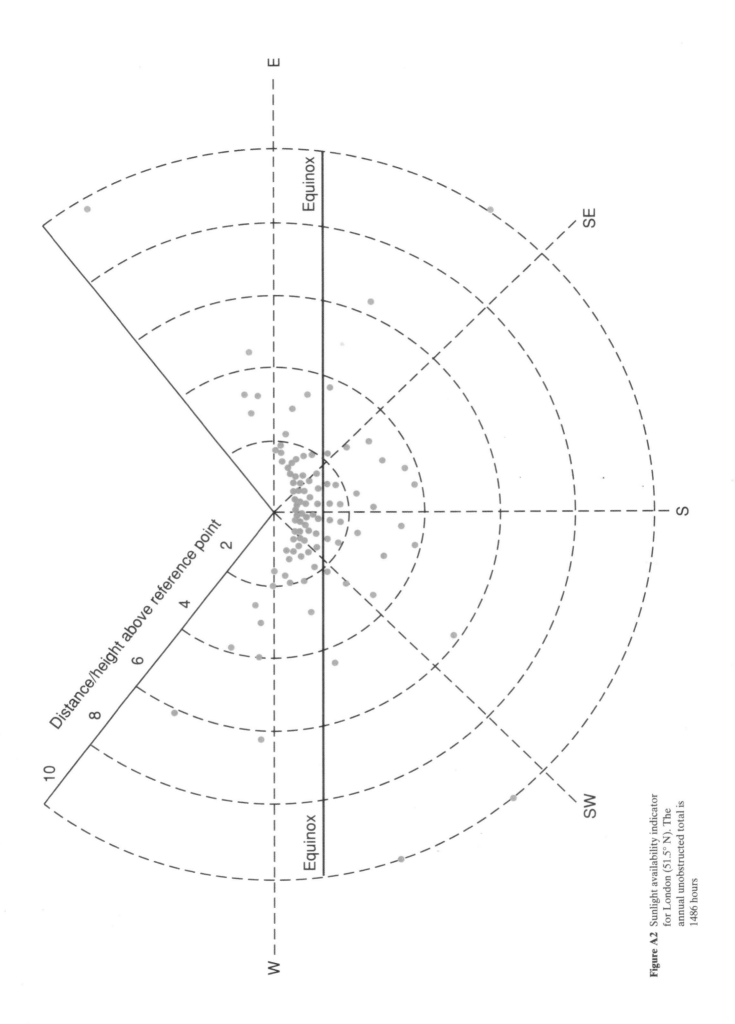

Figure A2 Sunlight availability indicator for London (51.5° N). The annual unobstructed total is 1486 hours

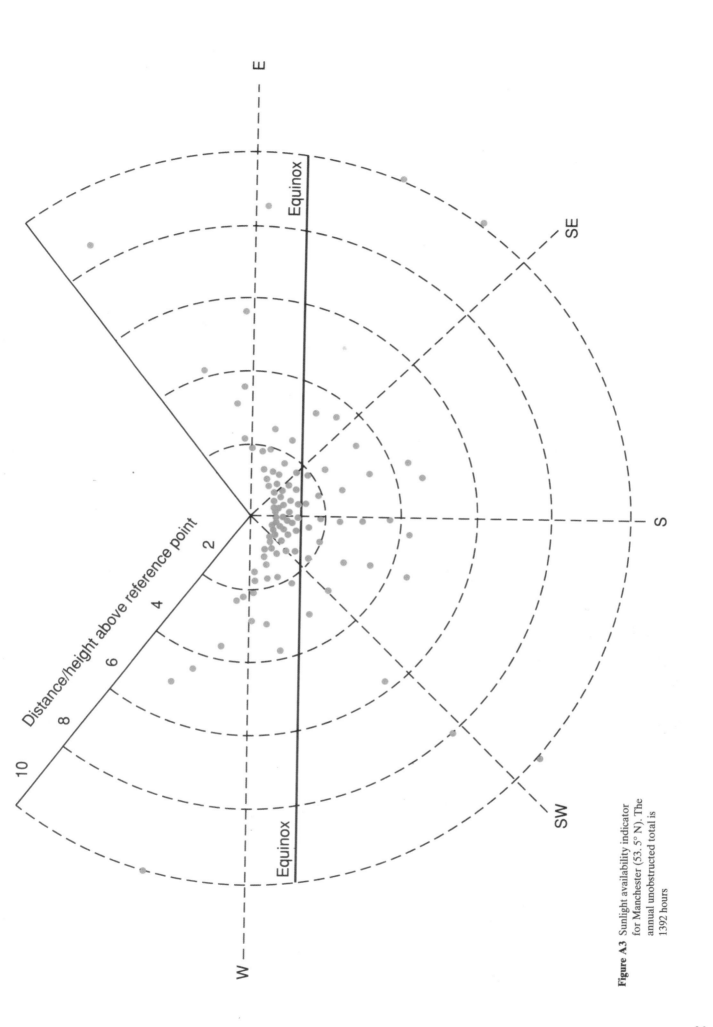

Figure A3 Sunlight availability indicator for Manchester (53. 5° N). The annual unobstructed total is 1392 hours

29

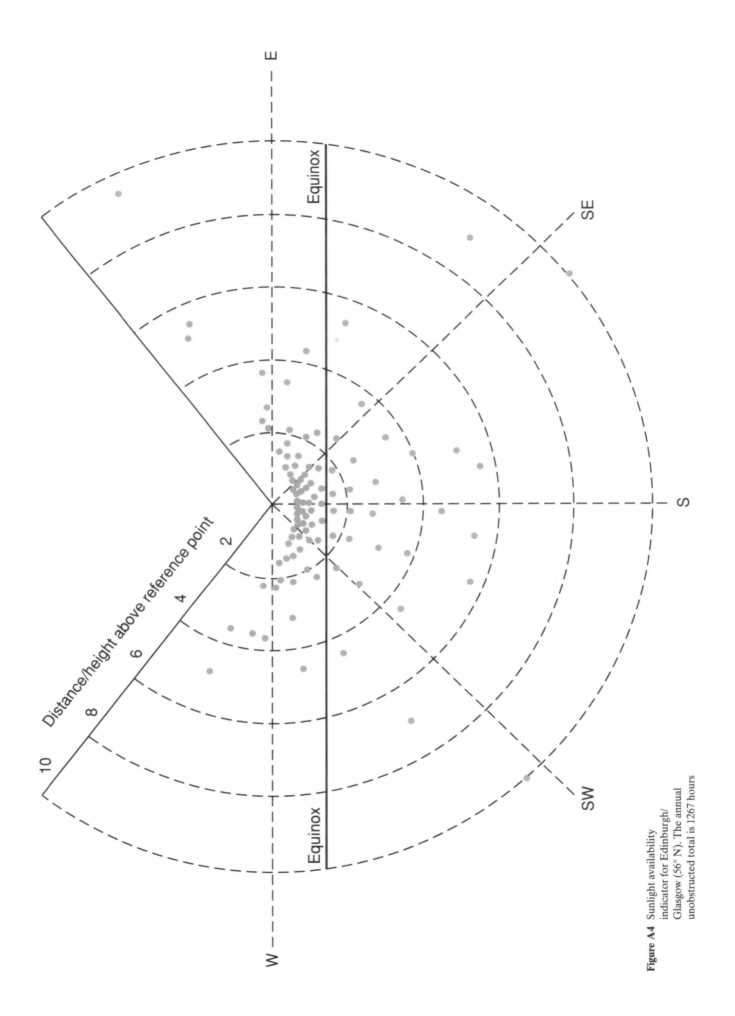

Figure A4 Sunlight availability indicator for Edinburgh/ Glasgow (56° N). The annual unobstructed total is 1267 hours

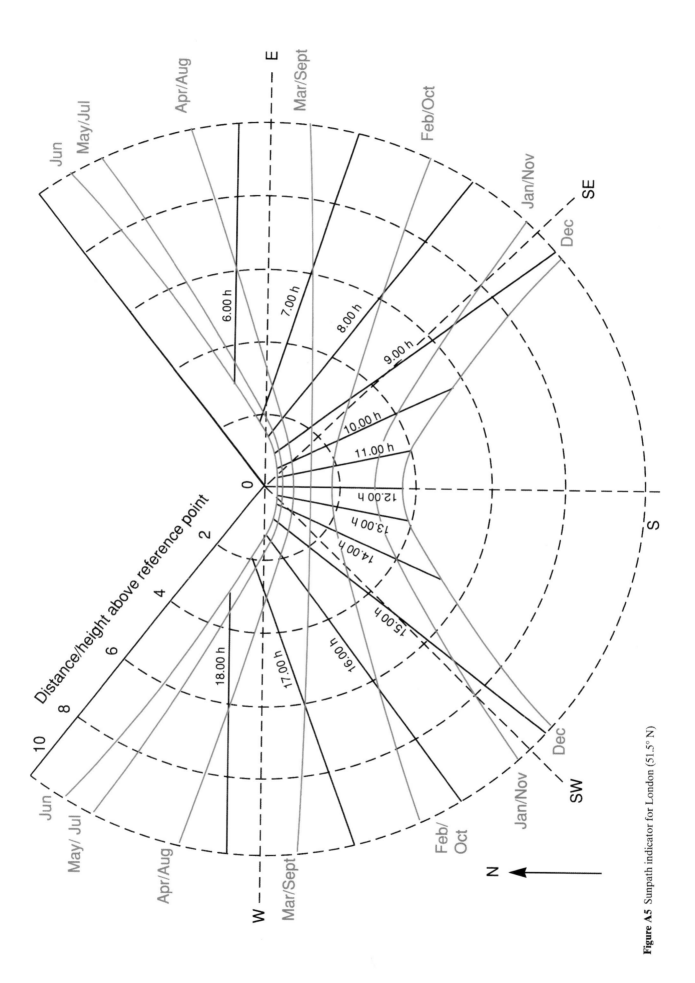

Figure A5 Sunpath indicator for London (51.5° N)

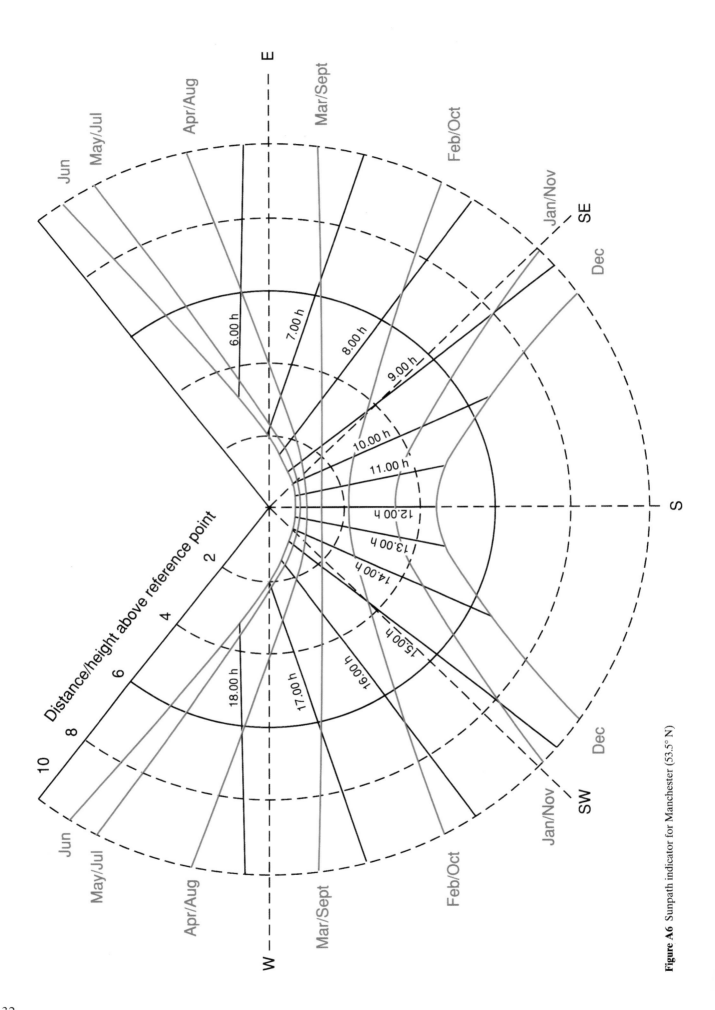

Figure A6 Sunpath indicator for Manchester (53.5° N)

32

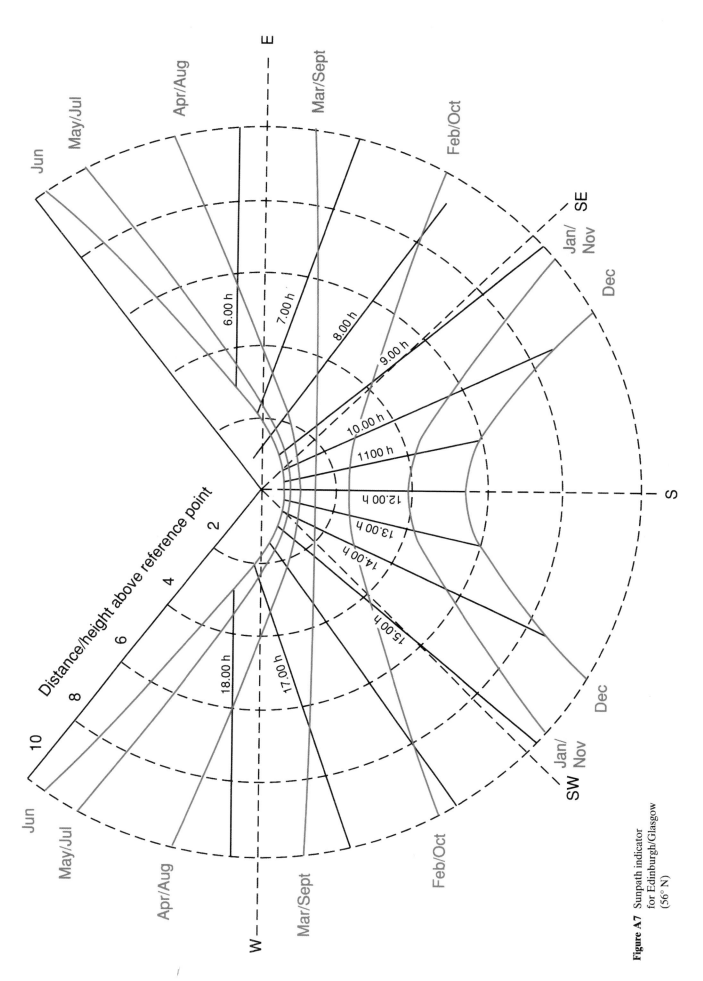

Figure A7 Sunpath indicator for Edinburgh/Glasgow (56° N)

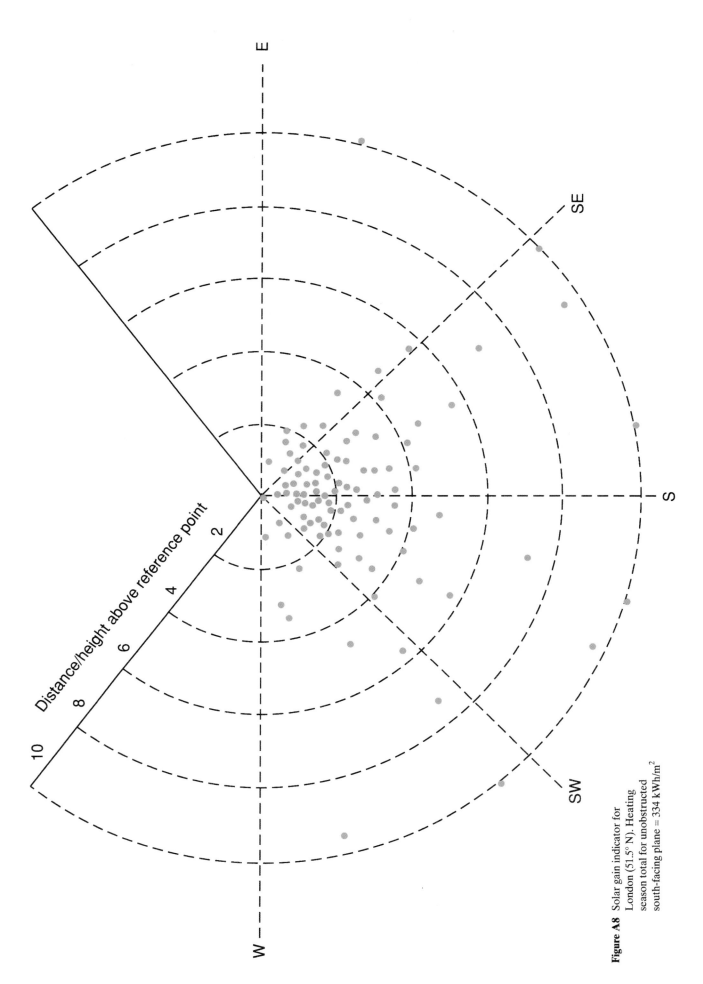

Figure A8 Solar gain indicator for London (51.5° N). Heating season total for unobstructed south-facing plane = 334 kWh/m²

Distance/height above reference point

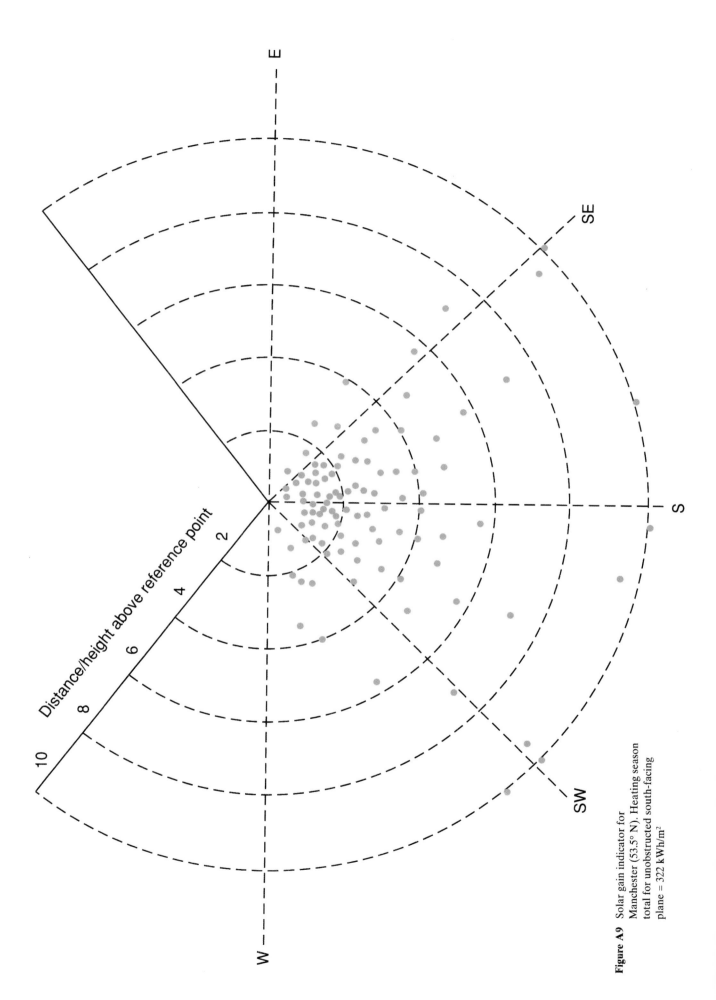

Figure A9 Solar gain indicator for Manchester (53.5° N). Heating season total for unobstructed south-facing plane = 322 kWh/m²

35

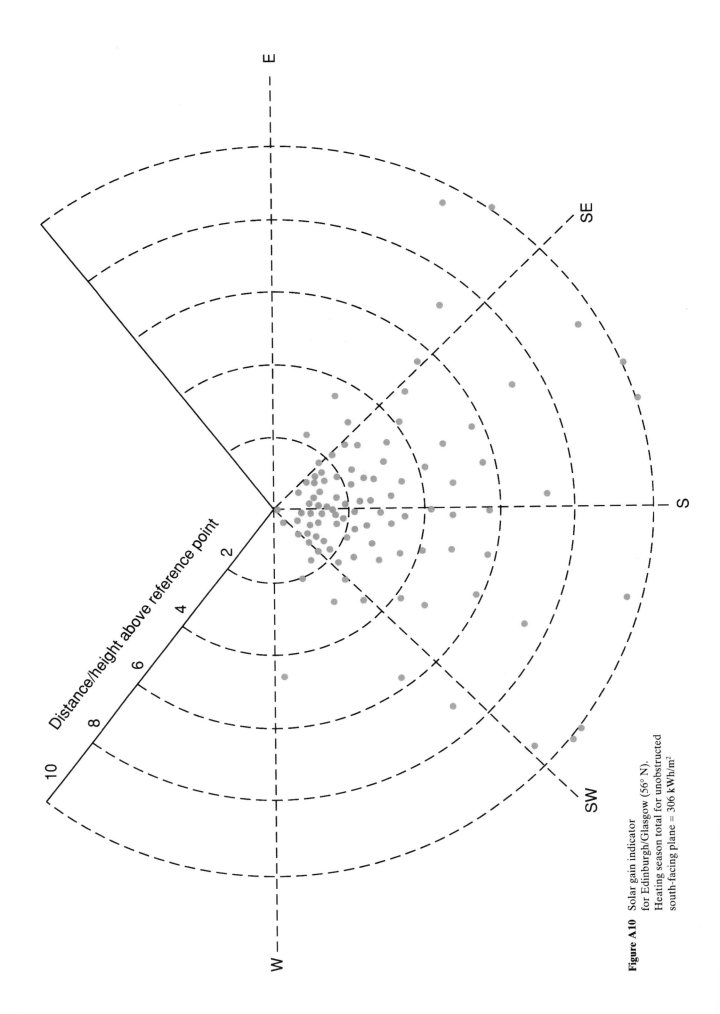

Figure A10 Solar gain indicator for Edinburgh/Glasgow (56° N). Heating season total for unobstructed south-facing plane = 306 kWh/m²

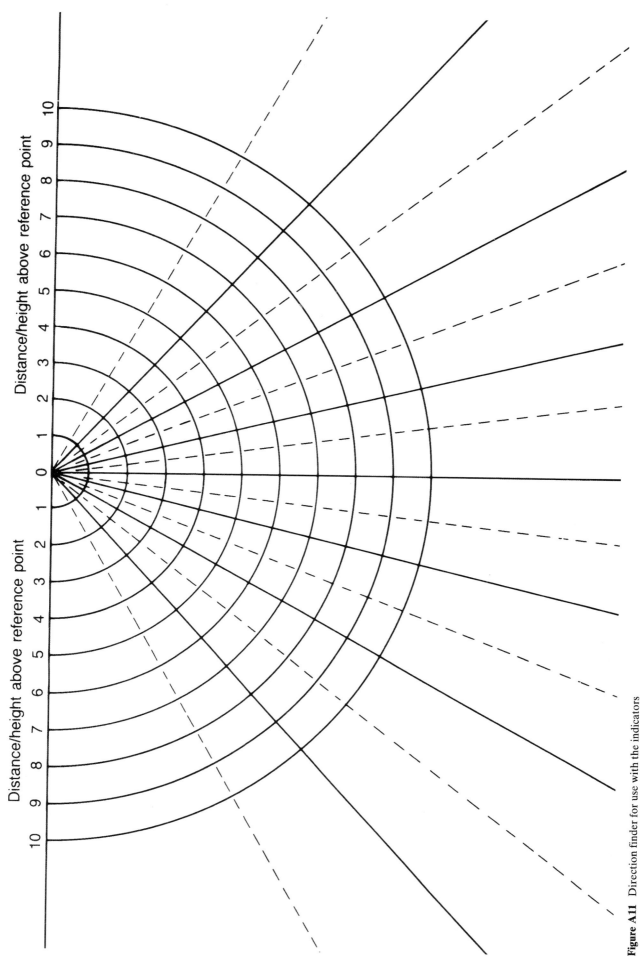

Figure A11 Direction finder for use with the indicators

37

Figure A12 gives a plan of an example housing layout. You wish to find the daylight and sunlight reaching point O on the plan. Mark the layout on the direction finder, as shown in Figure A13. Lay the base of the direction finder parallel to face POQ. Then consider each radial zone in turn. Shed ABCD blocks about half of the far left zone. Its nearest face is on average 12 m from O and it is 1 m above point O, so its distance-to-height ratio is 12. This is actually off the scale of the direction finder, so this shed is unlikely to be a significant obstruction to daylight. In general, obstructions whose distance is more than 10 times their relative height can be ignored.

EFGH is a house with a pitched roof. It blocks a full radial zone and two halves. In the full zone, the eaves EF are 20 m from O and 5 m above it, so the distance-to-height ratio is 4. The ridge IJ is 24 m from O and 8 m above it; so its ratio is 3. In this case, then, the ridge forms the highest part of the obstruction as seen from O, because its distance-to-height ratio is least. This radial zone is therefore marked off at the value 3 on the radial scale (Figure A13). The two half-zones are marked off in a similar way.

Extension QR covers two zones at an oblique angle. In this situation it is usually easier and more accurate to plot the ends Q and R of the obstruction on the direction finder, and join them up. These have been marked as Q' and R'. Point R, for example is 10 m on plan from O and 5 m above it. So R' is marked where the radial line QR intersects the distance–height arc of 2 (10/5). Note that any horizontal edge plotted in this way on the direction finder should be parallel to the same horizontal edge on the original plan. This can be used as a general way of plotting horizontal edges.

Only part of house KLMN is plotted on the direction finder. This is because face KL lies behind and below extension QR as seen from O. Only the side of the house KN need be plotted.

The resulting plot of the obstructions can then be used with the indicators as described next. For use with the sunpath, sunlight availability, and solar gain indicators, the south point of the layout should be marked on the direction finder.

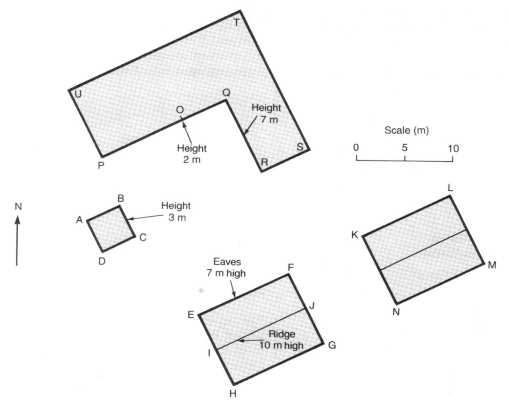

Figure A12 Site plan of an example situation

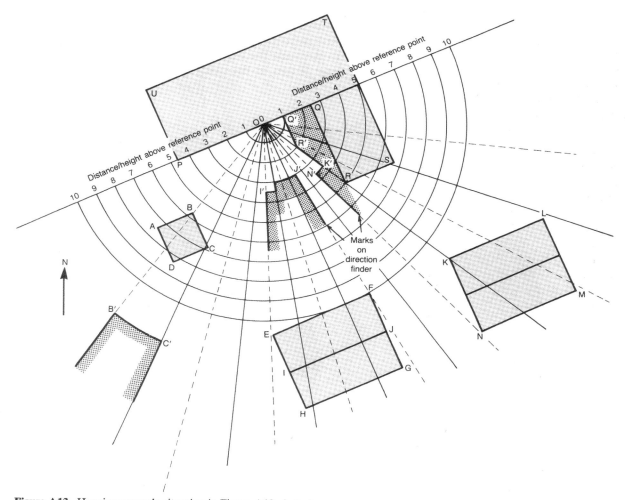

Figure A13 Housing example situation in Figure A12 plotted on
to direction finder

Use of a plan drawn to a specific scale

If there are only one or two obstructions, or if all the obstructions are the same height above the reference point, it is usually easier not to use the direction finder but to draw a special plan. The plan is drawn on tracing paper so that it can be laid over the indicators.

It is essential to draw the plan to the correct scale. The scale to be used will depend on the height of the obstruction above the reference point for the calculation. If this distance is h m then the plan should be drawn to a scale of $1:100h$. Table A1 gives scales for some obstruction heights

Table A1 Scales of plan for use directly with indicators

Height h of obstruction above reference point (m)	Scale of plan
1	1:100
2	1:200
3	1:300
4	1:400
5	1:500
6	1:600
7	1:700
8	1:800
9	1:900
10	1:1000
15	1:1500
20	1:2000

Figure A14 illustrates an example. It is a plan of an office block 10 m high. This has a central, three-sided courtyard. The daylight and sunlight reaching point O, 2 m above ground on the west side of the courtyard, need to be found.

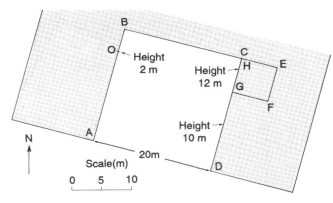

Figure A14 Site plan of an example situation

The top of the courtyard is 8 m above point O, so the plan of the building needs to be redrawn to a scale 1:800 (Figure A15). The width of the courtyard (20 m) will be 25 mm in this scale.

If some obstructions are of a different height they can be included by superimposing a plan, drawn to a different scale, onto the first one. For example, in Figure A14, EFGH is a plant room 10 m above point O. It is drawn to a scale of 1:1000 on the plan, with point O remaining at the same place as before. In this scale the effective width of the courtyard is now 20 mm, so the plant room appears nearer to O.

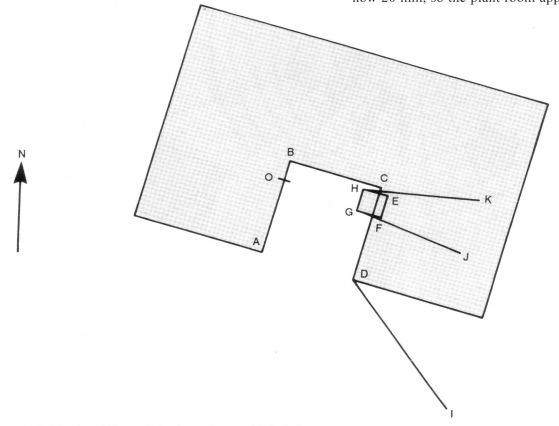

Figure A15 Site plan of Figure A14 redrawn for use with the indicators

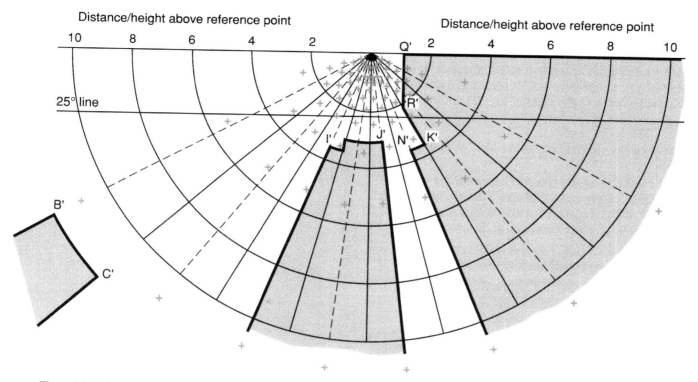

25° line

Figure A16 Direction finder plot of housing layout (Figures A12, A13) laid over skylight indicator

If the daylight and sunlight are required on a vertical facade (in this case BOA), the line of the facade should be drawn in (it does not matter if it is a different height to the obstructions). For sunlight calculations, a south point should also be included. Finally, mark out those areas of the plan from which point O is prevented from receiving light. Do this by drawing radial lines outwards from O from each end of the obstruction. Lines DI, GJ and HK are examples. Then the plan can be laid over the indicators as follows.

A2   Use of the skylight indicator

Take the marked direction finder or specially prepared plan and lay it over the skylight indicator (Figure A1). The centre of the indicator should correspond to the reference point, with its base parallel to the line of the vertical plane.

The skylight indicator has 80 crosses marked on it. Each of these corresponds to 0.5% vertical sky component. If a cross lies nearer to the centre of the indicator than any obstruction in that direction (as marked on the direction finder or special site plan), then it is unobstructed and counts towards the total vertical sky component. If it lies beyond the obstruction then it will be obstructed and does not count. The vertical sky component at the reference point (in %) is found by counting the unobstructed crosses and dividing by two. If a cross lies on the edge of a plotted obstruction, half a cross can be counted.

Figure A16 shows how vertical sky component is found in the housing layout example (Figure A12). The marked direction finder (Figure A13) is laid over the skylight indicator. For clarity, some of the numbers and semicircles on the direction finder have been omitted. The shaded areas lie beyond the obstructions and therefore crosses in these areas do not count. That leaves 62 ½ crosses in the unshaded areas, which will contribute to the skylight at point O; a vertical sky component of just over 31%.

In Figure A17 the courtyard layout (Figure A14) is analysed. The specially drawn site plan (Figure A15), on tracing paper, is laid over the skylight indicator. Here the number of unobstructed crosses is 46, so the vertical sky component is 23%. Note that crosses in the shaded area to the right of line DI do not count, as these areas of sky would be blocked by face DC.

Counting the crosses can be speeded up in various ways.

- If a point is lightly obstructed, count the obstructed crosses, divide by two and subtract this from 40% (the value for an unobstructed vertical plane) to give the vertical sky component.

- Each of the 8 radial zones contains 10 crosses, so if a zone is completely unobstructed it will contribute 5% to the vertical sky component.

- To check whether the vertical sky component exceeds 27% (the value for a long, straight obstruction subtending 25° on section), the 25° line can be used. If the number of obstructed crosses above the 25° line (between it and the reference point) is lower than the number of unobstructed crosses below this line then the 27% value is exceeded. In Figure A17 there are 13½ obstructed crosses above the 25° line, and only 5½ unobstructed crosses below it. So the courtyard layout has a vertical sky component of less than 27%.

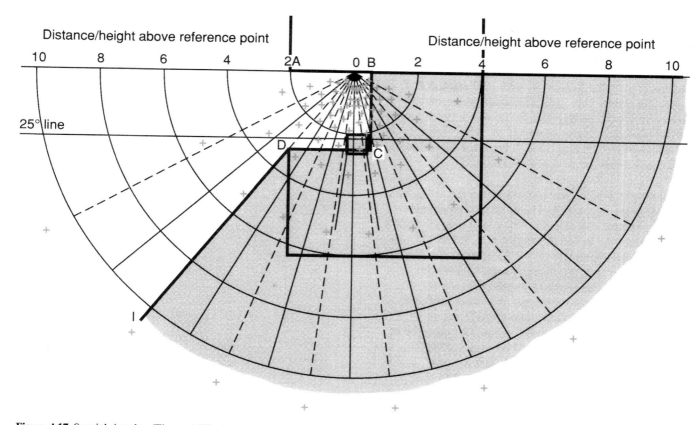

Figure A17 Special site plan (Figure A15) of courtyard layout laid over skylight indicator

A3 Use of the sunlight availability indicators

To estimate sunlight availability, first choose the indicator (Figures A2 to A4) which corresponds best to the latitude of the site (see Section A1). Take the marked direction finder or specially prepared plan and lay it over the indicator, with its centre at the reference point. This time, the south point on the indicator should be parallel to the south point marked on the direction finder or site plan, regardless of which way the building faces. This is an important difference from the skylight indicator.

The sunlight availability indicator has 100 spots on it. Each of these represents 1% of annual probable sunlight hours. The effects of obstructions are found in the same way as for the skylight indicator. If a spot lies nearer to the centre of the indicator than any obstruction in that direction (as marked on the direction finder or special site plan) then it is unobstructed. The percentage of annual probable sunlight hours at the reference point is found by counting all the unobstructed spots. The number of annual sunlight hours is found by multiplying this percentage by the annual total hours for an unobstructed plane, given in the caption to the indicator.

The British Standard[1] recommends that at least 25% of annual probable sunlight hours be available at the reference point, including at least 5% of annual probable sunlight hours in the winter months, between September 21 and March 21. To check this, use the horizontal equinox line on the indicator. At least 25 spots should be unobstructed, with 5 or more of them below the equinox line.

If the calculation point is on a wall of a building, then sunlight from behind the building cannot of course reach it, so any spots on the building side of the wall on which the reference point lies will be obstructed.

Figure A18 shows this indicator being used for the housing layout in Figure A12. The marked direction finder (Figure A13) is laid over the indicator (for London 51.5° N — Figure A2). Its south direction, taken originally from the site plan, is parallel to the south point on the indicator. For clarity, the markings and semi-circles on the direction finder are not shown. The shaded areas would be blocked by the obstructions. Note that areas above the direction finder base are also shaded, because sunlight from these directions would be blocked by the building on which point O lies.

From Figure A18, the total percentage of annual probable sunlight hours reaching point O is 62%, of which 22% are in the six winter months. Thus, point O easily meets the British Standard[1] recommendation. The total probable sunlight hours reaching O is 62% of 1486, or 921 hours a year.

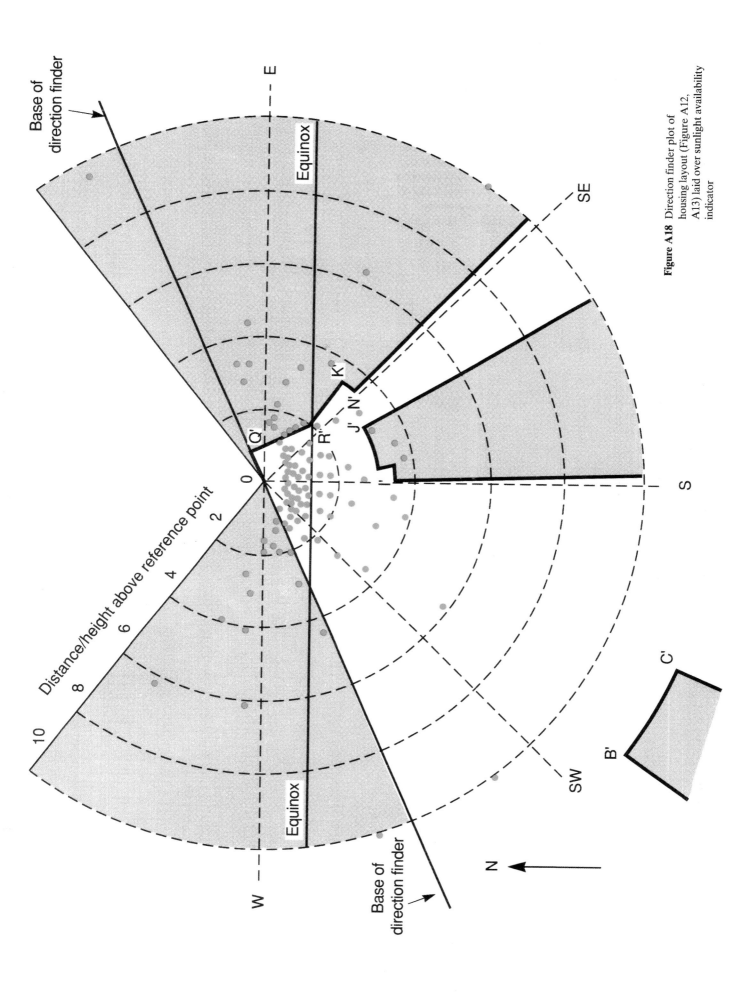

Figure A18 Direction finder plot of housing layout (Figure A12, A13) laid over sunlight availability indicator

In Figure A19 the courtyard layout (Figure A14) is assessed. The specially drawn site plan (Figure A15) is laid over the indicator, with the two south directions, on plan and indicator, parallel. Note that wall BOA has been continued to A' because sunlight from areas to the north west of AA' will not be able to reach point O. Similarly, sunlight from north east of line DI will be prevented by wall CD from reaching point O. The total sunlight availability at O is 48% of annual probable hours, 16% of these being in the winter months. Thus, the British Standard[1] is satisfied here too.

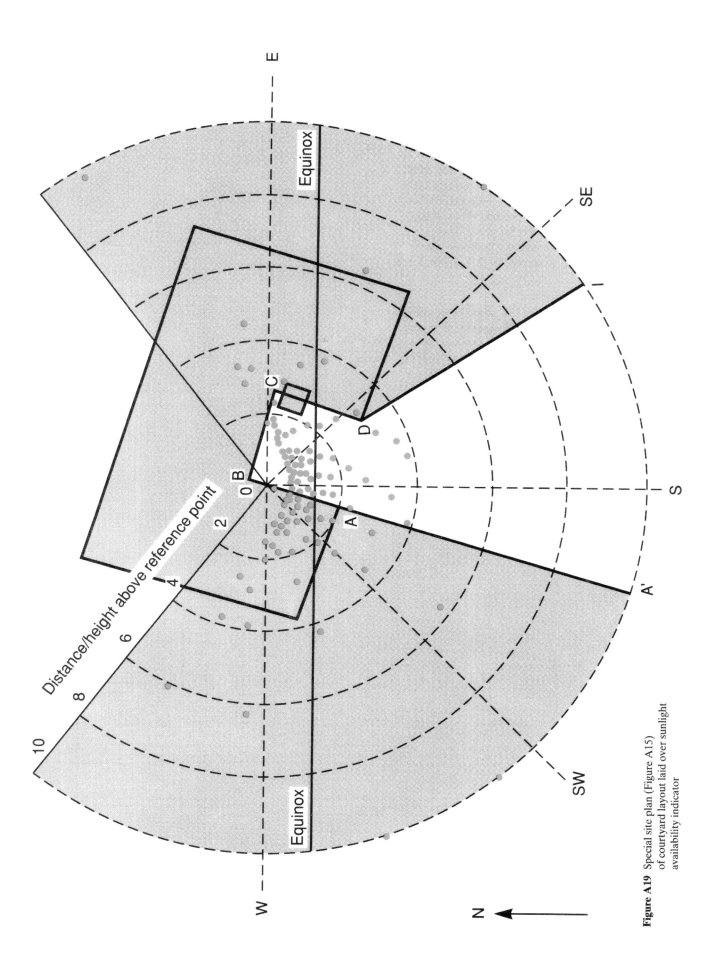

Figure A19 Special site plan (Figure A15) of courtyard layout laid over sunlight availability indicator

47

A4 Use of the sunpath indicators

The sunpath indicators (Figures A5 to A7) give the times of day and year at which sunlight can reach a reference point. The bold curved lines which run across the indicator are sunpaths for approximately the 21st of each month (the months are labelled around the perimeter). Each sunpath is divided into hours by the thinner, straight, solid hour-lines which radiate outwards. These are labelled with solar time, which is almost the same as Greenwich Mean Time (GMT); add one hour for British Summer Time. Where the time of day is unusually important, solar time can be corrected using the method described in a BRE Report[2].

The sunpath indicator is used in the same basic way as the sunlight availability indicator. Figure A20 shows the housing layout plot (Figure A13) superimposed on the London sunpath indicator. For April, August, May, July and June, sunlight will be available at point O from around 08.30 h GMT, when the sun appears above wall QR, until just after 15.00 h, when it goes round past point P in Figure A12. This corresponds to 09.30 h until 16.00h British Summer Time. In February, October, March and September, there is sun from 09.00h GMT until 1600h. In December, January and November, there is sun from around 09.00 h until 10.00 h. From 10.00h to 12.00h building IJ blocks it, then from 12.00 h onwards the sun is unobstructed until it sets.

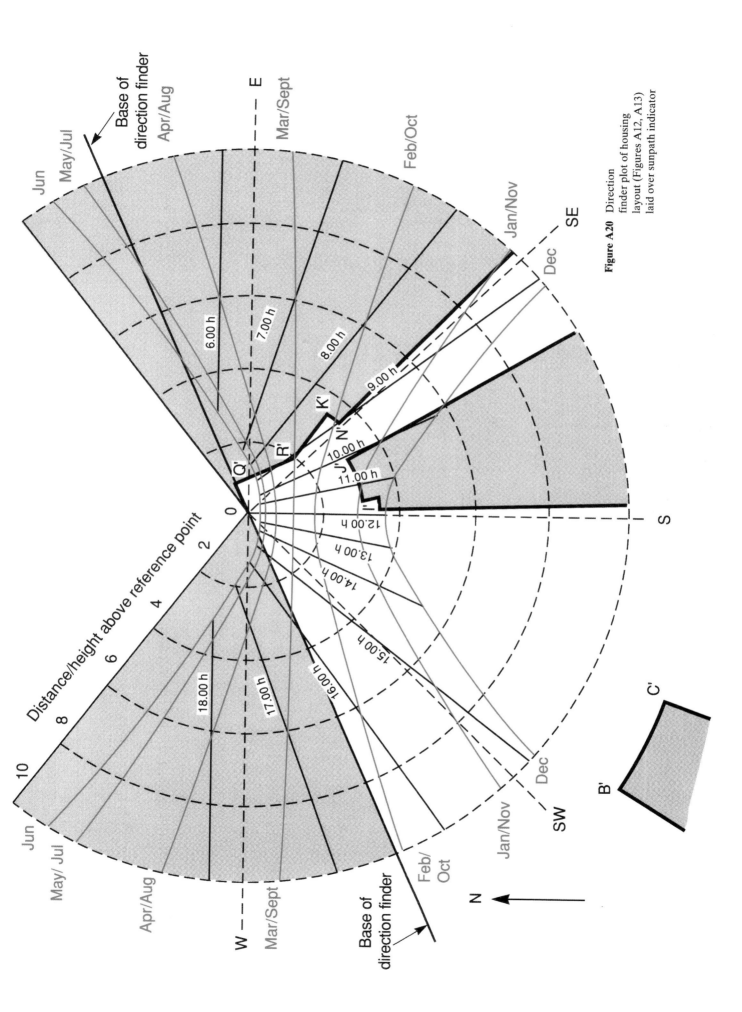

Figure A20 Direction finder plot of housing layout (Figures A12, A13) laid over sunpath indicator

49

Figure A21 shows this indicator with the special plan of the courtyard layout (Figure A15). In April, August, May, July and June, sunlight is available from around 07.30h GMT, when the sun appears over wall BC, until just after 12.30h, when it no longer can shine on face BOA. In March and September there is sun from 08.15h GMT until 13.00h, and in February and October from 09.15h until 13.00h. In January, November and December, there is sun from 09.45h, when it appears around end D of wall CD, until 13.00h.

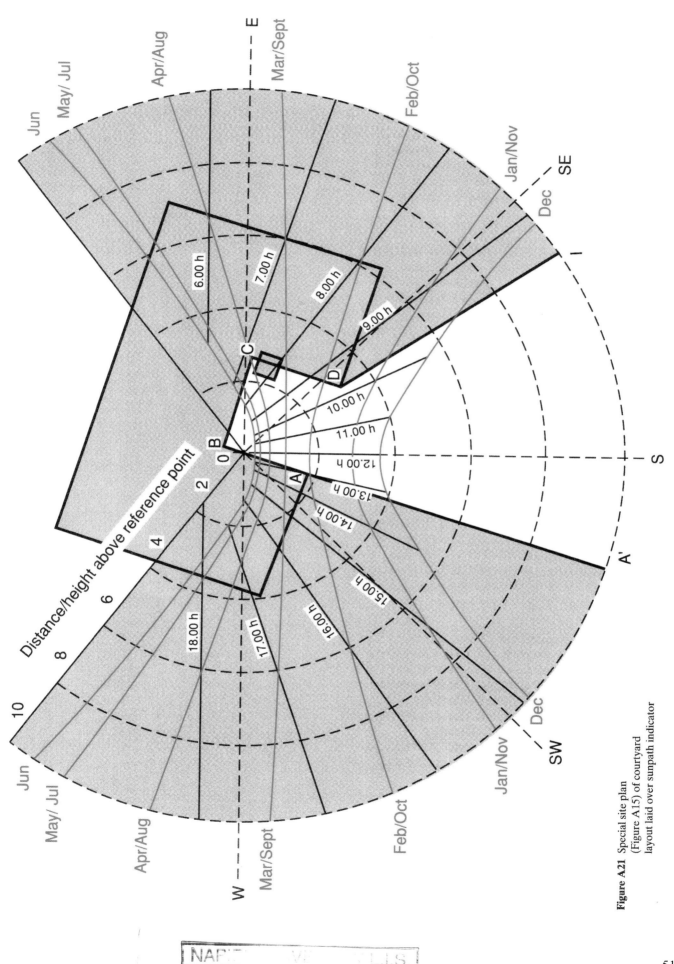

Figure A21 Special site plan (Figure A15) of courtyard layout laid over sunpath indicator

51

A5 Use of the solar gain indicators

The solar gain indicators (Figures A8 to A10) can be used to find the incident solar radiation on a south-facing vertical wall or window, summed over the heating season (October to April inclusive). They do not include summer solar gain, so cannot be used as a check on overheating. They are to help to estimate the contribution of solar radiation to space heating. No correction is made for glass losses or for whether the incident solar gain is useful. Both diffuse sky and direct solar radiation are included but not ground reflected radiation.

The indicator is calculated for a south-facing wall, but will give acceptable results for any vertical plane facing within 30° of due south. It should not be used for walls facing further east or west than this, or for sloping surfaces such as roof solar collectors.

The solar gain indicator is used in exactly the same way as the sunlight availability indicator (Section A3). The only difference is that each unobstructed spot represents 1% of incident heating season solar radiation. The total value, in kWh/m^2, for an unobstructed plane is given in the caption for each indicator.

Figure A22 shows the direction finder plot (Figure A13) of the housing layout (Figure A12) superimposed on the solar gain indicator. As before, the south direction taken from the site plan should be parallel to the south point on the indicator, whether the wall faces exactly due south or not. If the wall does not face due south, then spots behind the line of the wall (ie, the base of the direction finder) should be discounted.

The total number of unobstructed spots is then counted. For a lightly obstructed, passive solar layout it may be easier to count the obstructed spots and subtract from 100. In the housing layout 82 spots are unobstructed, so 82% of total available solar radiation during the heating season reaches point O. This amounts to 334 x 82% = 274 kWh/m^2. Point O on the courtyard layout (Figures A14 and A15) cannot be analysed with this indicator because its wall faces more than 30° from due south.

References to Appendix A

1 **British Standards Institution.** Code of practice for daylighting. *British Standard* BS 8206 Part 2: 1992. London, BSI, 1992.

2 **Hunt D R G.** *Availability of daylight.* Building Research Establishment Report. Garston, BRE, 1979.

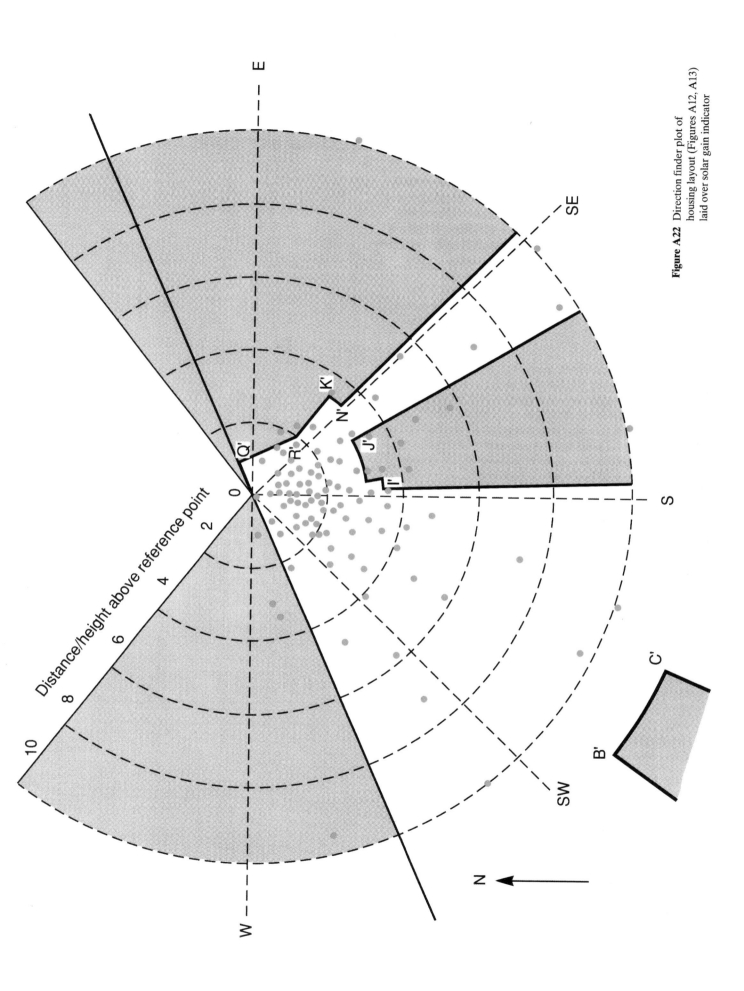

Figure A22 Direction finder plot of housing layout (Figures A12, A13) laid over solar gain indicator

Distance/height above reference point

53

Appendix B

A Waldram diagram to calculate vertical sky component

As an alternative to the skylight indicator described in Appendix A, a special form of Waldram diagram (Figure B1) can be used to estimate the vertical sky component on an external wall or window. Although it will usually be more time consuming to use than the skylight indicator, the Waldram diagram is more precise, and may be used for very complex obstructions. The basic approach is to plot all the obstructions on the diagram; the remaining area is proportional to the sky component on the vertical plane.

Figure B1 is used in the same way as the conventional Waldram diagram for interior daylighting, except that no window outline needs to be plotted, as only external surfaces are being considered. Each square centimetre on the diagram corresponds to 0.1% of sky component. Its total area is just under 400 cm^2, corresponding to the sky component of just under 40% on an unobstructed vertical plane.

The horizontal scale on the diagram is the azimuth angle in degrees from the line perpendicular to the vertical reference plane. The vertical scale is the altitude angle in degrees above the horizontal measured from the reference point on the vertical plane (usually 2 m from ground level). On the diagram, vertical edges of obstructions plot as straight vertical lines; horizontal or sloping edges generally plot as curved lines.

To plot a corner of an obstruction or a point on a sloping edge, first measure the angle on the plan at the reference point between the line to the point on the obstruction and the perpendicular to the window wall. This gives the position on the azimuth scale of the diagram. The position on the altitude scale is given by:

$$\text{Altitude angle} = \arctan(h/d) \text{ degrees}$$

where h is the height of the point on the obstruction above the reference point, and d is the distance between the two points on plan. In this case, the centre scale of the diagram should be used, ignoring the droop lines. This altitude angle is not necessarily the same as the angle on any sectional drawing...

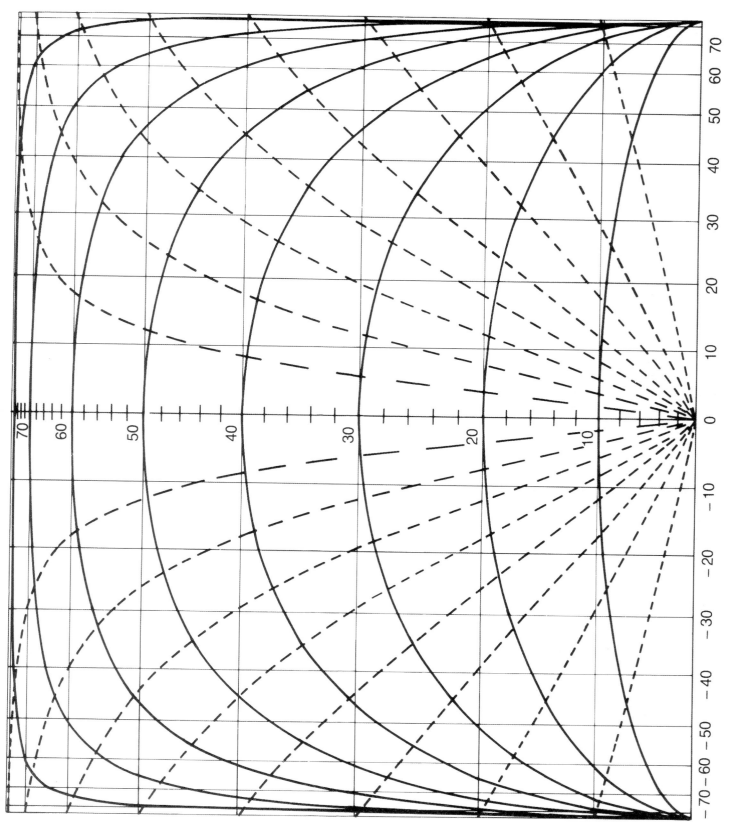

Figure B1 Waldram diagram for calculating vertical sky component

... for example, suppose point B on the roof line of the obstructing building in Figure B2 needs to be plotted. Its azimuth angle measured from the plan is 40°. On plan it is 20 m from the reference point P and it is 8 m above it on section. So its altitude angle is arctan (8/20) = arctan (0.4) = 22°. These two angles give its co-ordinates on the Waldram diagram (Figure B3).

The droop lines on the diagram can be used to plot horizontal edges. The solid droop lines are for edges parallel to the plane of the reference point. The droop line is chosen according to the altitude angle of the horizontal edge in a section perpendicular to the reference window wall. So, for example, in Figure B2 the altitude of the ridge line CD is 30°. It is therefore plotted (Figure B3) along the 30° solid droop line, between azimuth angles corresponding to those of C and D on the plan.

The broken droop lines on the diagram are used to plot horizontal edges perpendicular to the plane of the reference point. The side FG of the roof of the extension in Figure B2 can be plotted in this way. The required droop line can be chosen by finding the co-ordinates of any point on the obstructing edge using the method just described. Alternatively, if an elevation of the wall containing the reference point is available, the angular altitude of the horizontal edge can be measured off it. The correct droop line is the one which intersects the side of the diagram at that point on the altitude scale. In our example, point G has an altitude of 20° measured on the elevation (Figure B2(c)). It is plotted at the far edge of the Waldram diagram (Figure B3) at 20° on the altitude scale. The broken droop line through this point is the edge of the extension FG.

Once all the obstructions have been plotted, measure the remaining area not covered by obstructions (squared tracing paper is ideal for this). Then divide it by 10 to get the vertical sky component. In our example, the unobstructed area on the diagram (Figure B3) is just over 290 cm², so the vertical sky component is just over 29%.

(a) **Section**

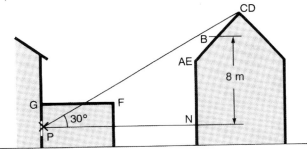

(b) **Plan**

(c) **Elevation**

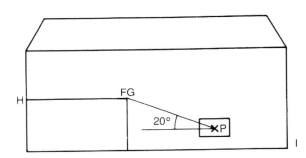

Figure B2 Section, plan and elevation of a (hypothetical) example situation

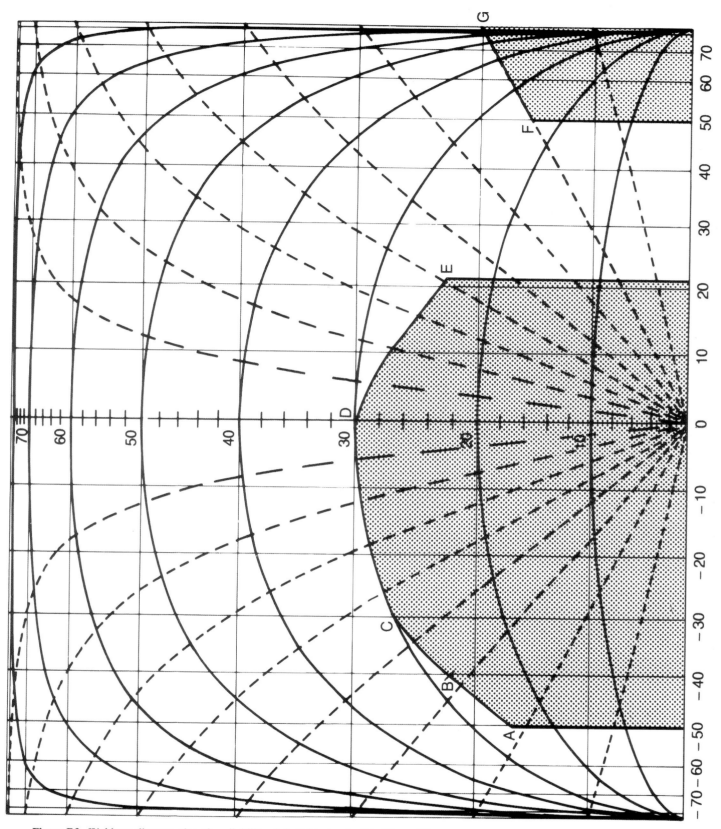

Figure B3 Waldram diagram plot of availability of skylight at point P in Figure B2

Appendix C

Interior daylighting recommendations

The British Standard for daylighting[1], and the CIBSE *Applications manual: window design*[2], contain advice and guidance on interior daylighting. This guide to good practice is intended to be used in conjunction with them, and its guidance is intended to fit in with their recommendations.

For skylight, the British Standard[1] and the CIBSE manual[2] put forward three main criteria, based on the average daylight factor, room depth, and the position of the no-sky line.

Average daylight factor (*df*)

This is defined in Figure C1. If a predominantly daylit appearance is required, then *df* should be 5% or more if there is no supplementary electric lighting, or 2% or more if supplementary electric lighting is provided. There are additional recommendations for dwellings, of 2% for kitchens, 1.5% for living rooms and 1% for bedrooms. These last are minimum values of average daylight factor, which should be attained even if a predominantly daylit appearance is not required.

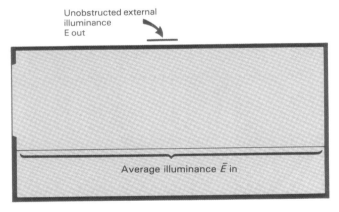

Figure C1 Definition of average daylight factor
Under standard overcast conditions:
Average daylight factor $df = \dfrac{\bar{E}\ \text{in}}{E\ \text{out}} \times 100\%$

The average daylight factor can be calculated using the following formula:

$$df = \frac{TA_{w}\theta}{A(1-R^2)}\ \%$$

where

T is the diffuse visible transmittance of the glazing, including corrections for dirt on glass and any blinds or curtains (For clean clear single glass, a value of 0.8 can be used.)

A_{w} is the net glazed area of the window (m²)

A is the total area of the room surfaces: ceiling, floor, walls and windows (m²)

R is their average reflectance. For fairly light-coloured rooms a value of 0.5 can be taken

θ is the angle of visible sky in degrees, measured as shown in Figure C2

Of these quantities, only θ depends on external obstruction. It can be directly related both to the obstruction angle and to the vertical sky component on the external window wall, as Table C1 shows. In the table, no correction has been made for light blocked by the window reveals.

So, whatever the shape of any obstruction, it is possible to calculate the vertical sky component at the centre of the window, using the skylight indicator (Appendix A) or the Waldram Diagram (Appendix B) to find the 'equivalent θ' from Table C1. This value can be used in the equation to find the average daylight factor for complex obstructions.

Reductions in the vertical sky component received by the windows of an existing building can also be related to reductions in *df* using Table C1. For

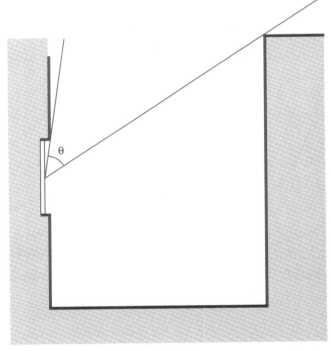

Figure C2 θ is the angle subtended, in the vertical plane normal to the window, by sky visible from the centre of the window

example, if the vertical sky component is reduced from 30% to 24% (0.8 times its former value), the value of θ will be reduced from 70 to 60. Thus, the *df* is reduced to 60/70 or 0.86 times its former value.

Table C1   Values of angle θ for obstruction angles and vertical sky components

Obstruction angle (° from horizontal)	Vertical sky component at centre of window (%)	Value of θ in average daylight factor equation
0	40	90
5	38	85
10	35	80
15	33	75
20	30	70
25	27	65
30	24	60
35	21	55
40	18	50
45	15	45
50	13	40
55	10	35

Room depth

If a daylit room is lit by windows in one wall only, the depth of the room L should not exceed the limiting value given by:

$$\frac{L}{W} + \frac{L}{H} \le \frac{2}{1 - R_b}$$

where

W is the room width
H is the window-head height above floor level
R_b is the average reflectance of surfaces in the rear half of the room (away from the window)

If L exceeds this value, the rear half of the room will tend to look gloomy and supplementary electric lighting will be required.

External obstructions do not influence this recommendation. However, there are implications for the site layout, because the recommendation relates to the maximum depth of a building that can be satisfactorily daylit (just over twice the limiting depth L, from window wall to window wall).

Position of the no-sky line

If a significant area of the working plane lies beyond the no-sky line (ie, it receives no direct skylight), then the distribution of daylight in the room will look poor and supplementary electric lighting will be required. Appendix D gives guidance on how to plot the no-sky line.

Note that all three of the criteria need to be satisfied if the whole of a room is to look adequately daylit. Even if the amount of daylight in a room (given by the average daylight factor) is sufficient, the overall daylit appearance will be impaired if its distribution is poor.

For sunlight, follow the British Standard recommendation[1] as outlined in Sections 3.1 and 3.2.

References to Appendix C

1   **British Standards Institution.** Code of practice for daylighting. *British Standard* BS 8206 Part 2: 1992. London, BSI, 1992.

2   **Chartered Institution of Building Services Engineers.** *Applications manual: window design.* London, CIBSE, 1987.

Plotting the no-sky line

The no-sky line divides those areas of the working plane which can receive direct skylight, from those which cannot. It is important for two main reasons:

● To indicate how good the distribution of daylight is in a room. Areas beyond the no-sky line will generally look gloomy

● As an aid in determining whether rights to light may be infringed (Appendix E)

If the external obstructions already exist, it is possible to measure directly the position of the no-sky line in a room. This is best done using a vertical pole, such as a small camera tripod, and adjusting its height to that of the working plane. By moving the tripod about, and kneeling or sitting on the floor, and sighting through the top of the tripod to the window head, it is possible to find the exact places at which the last patches of sky disappear (Figure D1). If furniture or walls make this difficult, a small plane or convex mirror mounted on the top of the tripod can be used as shown in Figure D2. The mirror need not be exactly horizontal.

In most cases the position of the no-sky line has to be found from plans. Figures D3 to D7 illustrate some common cases. It is usually easiest to have both a plan and a section drawn up.

Long horizontal obstruction parallel to window (Figure D3)
The no-sky line is also parallel to the window, distant

Here
    h is the height of the window head above the working plane
    y is the height of the obstruction above the window head
    x its distance from the outside window wall.

If d is greater than the room depth, no part of the room lies beyond this no-sky line.

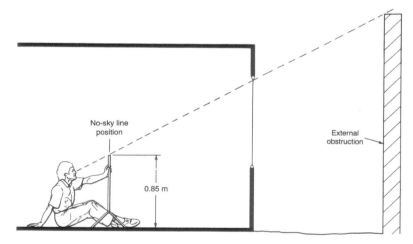

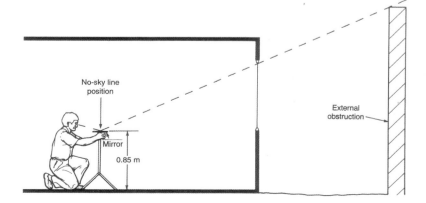

Figure D1 At the no-sky line, the last visible patch of sky above the obstructions will just disappear when the window head is sighted through a point at working plane height

Figure D2 A mirror can be used to sight the no-sky line (compare Figure D1)

Narrower horizontal obstruction parallel to window (Figure D4)

Here, the obstruction is the same height and distance away as in Figure D3, but it terminates at points A and B. CD is part of the same no-sky line as in Figure D3, but now points north of DE can receive light around corner A of the obstruction, and points south of CF can receive light around corner B. The no-sky area is in the form of a trapezium. If the obstruction AB had been even narrower, the no-sky

$$d = \frac{xh}{y} \quad \text{from its outside face.}$$

area would have been triangular in shape, and in the same position even if the obstruction were higher.

In plotting the no-sky line, the key points are the top corners of the window. These are usually the last points at which sky can be seen.

Horizontal obstruction perpendicular to window wall and projecting from it (Figure D5)

Part of the no-sky line (DB) runs parallel to the obstruction. Its distance d from the corner of the window A (the corner furthest from the obstruction) is again

Here

> h and y are the same heights as before
> x is the distance on plan from corner A to the obstruction measured along the window wall

The rest of the no-sky line BC is the continuation of the straight line FABC from the end of the obstruction. Points in the triangle EBC can receive skylight around the corner F; points in the triangle

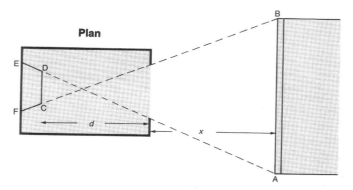

Plan

Figure D4 The no-sky area for a narrower obstruction is bounded by lines through the sides of the window and the vertical ends of the obstruction

ABD can 'see' sky over the top of the obstruction.

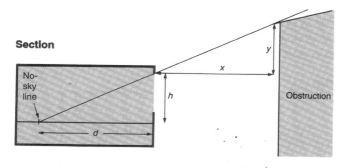

Section

No-sky line

Obstruction

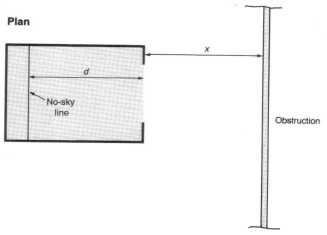

Plan

No-sky line

Obstruction

Figure D3 For a long obstruction parallel to the window, the no-sky line is also parallel to the window. Its position can be found from the section

This is a special case of a general rule: if there is a horizontal obstruction at any orientation relative to the window wall, then part of the no-sky line will be parallel to the obstruction. Its position is given by

where

$$d = \frac{xh}{y}$$

d and x are the perpendicular distances to the no-sky line and the obstruction, measured from the corner of the window furthest from the obstruction

h and y are the same as before (Figure D6). This

no-sky line may be curved rather than straight. For example, in Figure D5 the no-sky line BD would instead curve westwards as it neared the window wall, finally touching it at point A. This makes a significant difference only within about four or five wall

$$d = \frac{xh}{y}$$

thicknesses of the window wall.

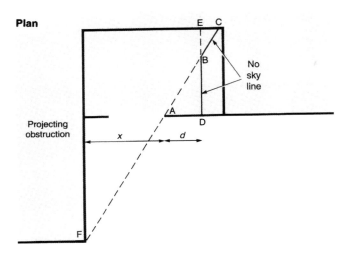

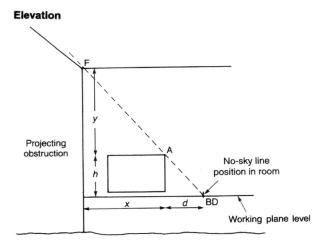

Figure D5   For an obstruction projecting from the window wall, the no-sky line runs partly parallel to the obstruction and partly along a continuation of the line joining the end of the obstruction F and the side of the window A

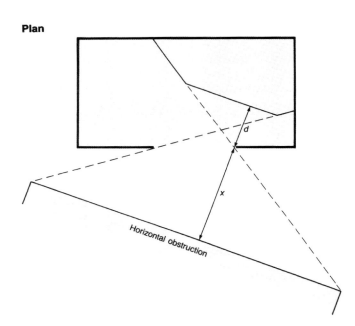

Figure D6   For a general horizontal obstruction, part of the no-sky line runs parallel to the obstruction. The rest runs along the projected lines from the ends of the obstruction through the corners of the window

applies unless the obstruction is too narrow (compare the case, already discussed, of the narrower horizontal obstruction parallel to window).

The analysis of Figures D5 and D6 assumes that the window wall is negligibly thin. If the window wall is thick, then the no-sky area is larger and part of the

More complex situations

These generally occur where there are a number of obstructions which, seen from the interior, appear to lie partly behind each other. A typical example is a row of detached or semi detached houses with gaps in between (Figure D7). The no-sky line PQRSTUVWX is drawn on the room plan. It is a combination of curved and straight lines. In area QRSTU sky is visible above the gap BC; in area VWXY it is visible above the gap DE.

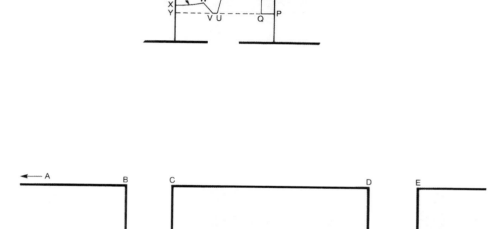

Plan

Figure D7 Complex no-sky line from a set of obstructions

In fact, in this particular case the no-sky line is a relatively poor indicator of the daylit appearance of the room. This is because only a small amount of skylight actually comes through the gaps BC and DE into the room. A better indicator of daylight distribution is, in fact, the line PY, which would be the no-sky line if ABCDE were a continuous terrace. In the particular case shown, no point in the regions QRSTU or VWXY will have a CIE sky component over 0.2% or a uniform sky factor over 0.25%, so line PY can be viewed as the effective no-sky line. It is, of course, easier to construct than PQRSTUVWX.

Where there is more than one window, the final no-sky line will surround those areas which cannot receive direct skylight from any of the windows. This can be arrived at by considering each window on its own at first, then combining them. In a room with windows on more than one side, it is often the case that all points on the working plane receive direct skylight through one window or another.

Appendix A

Rights to light

The right to light is a legal right which one property may acquire over the land of another. If a building or wall is erected which reduces the light in the obstructed property to below sufficient levels, then the right to light is infringed. The owner or tenant of the obstructed property may sue, either for removal of the obstruction or for damages. This can be costly if a whole building has to be pulled down, and so the question of rights to light should be considered at the design stage.

Rights to light can be acquired by a legal agreement, or if the light has been enjoyed without interruption for at least 20 years. If the light is obstructed for more than a year, then the right is usually lost. Sometimes, if windows have received light over adjoining land for nearly 20 years, the owner of the adjoining land may register a 'notional obstruction'. This is a way of stopping the windows acquiring rights to light over the land when the 20 years are up. Rights to light can also be rescinded by a legal agreement, usually with compensation to the owner of the property whose light is lost. No rights to light may be acquired over Crown land, although the windows of Crown buildings can acquire rights.

The right to light is for light from the sky alone. No right to sunlight exists — although there is a precedent for removal of obstructions to a greenhouse — neither is there a right to a view. Also, the right is to a bare minimum of light, in most circumstances well below what is recommended in the British Standard[1] (see Appendix C).

The usual way of calculating the loss of light is to compute the sky factor at a series of points on the working plane. In dwellings, the working plane height is usually taken to be 0.85m (33 inches). The sky factor is the ratio of the illuminance directly received from a uniform sky at the point indoors, to the illuminance outdoors under an unobstructed hemisphere of this sky. No allowance is made for glass losses or light blocked by glazed bars and (usually) window frames; nor is reflected light included, either from interior surfaces or from obstructions outside. The sky factor is, therefore, not the same as the CIE daylight factor (see Appendix C).

The sky factor is usually calculated using a Waldram diagram (Figure E1). This is used in the same way as the vertical surface Waldram diagram in Appendix B, except that the window needs to be plotted as well. The vertical edges of the window plot as straight vertical lines on the diagram, at the azimuth values measured on plan from the perpendicular from the calculation point to the window wall. The head and sill of the window are plotted along the solid droop line corresponding to their altitudes above the horizontal, measured in the section perpendicular to the window wall.

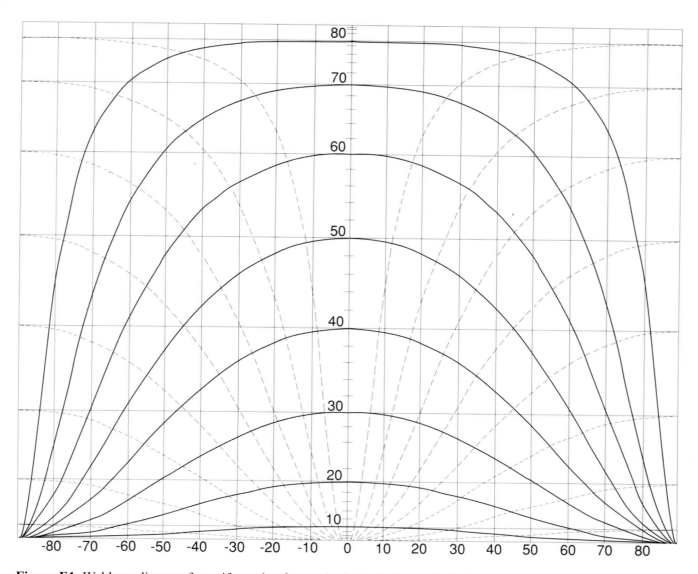

Figure E1 Waldram diagram for uniform sky, for use in rights-to-light calculations

65

Figure E2 gives an example, which is plotted as shown in Figure E3.

Obstructions are plotted as shown in Appendix B, all angles and distances being measured from the reference point inside the room. Only obstructions visible through the window need be plotted. Unless the point in the room lies beyond the no-sky line, the result on the diagram should be at least one patch of sky visible through the window and above the obstructions. The area of this patch (or patches) is proportional to the sky factor at the point. In Figure E1, each square centimetre represents 0.2% sky factor (the total area of the diagram is 250cm², corresponding to 50% sky factor). Note that the scaling factor in Figure E1 is different from the vertical Waldram diagram in Figure B1. Thus, in Figure E3 the area of the patch is 3.4 cm², so the sky factor is 0.68%. If there are windows in more than one wall, find the sky factors due to each, using separate Waldram diagrams, then simply add them together.

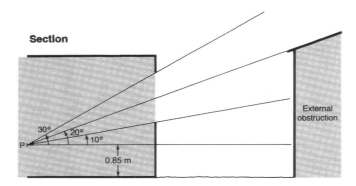

Section

External obstruction

30° 20° 10°

P

0.85 m

Plan

P

30° 10°

External obstruction

Figure E2 Plan and section of an example situation; room with wide, low nearby obstruction

66

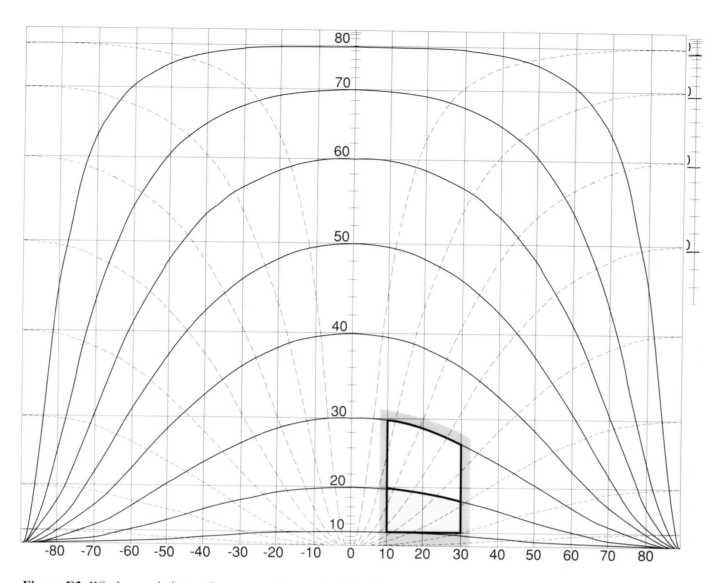

Figure E3 Window and obstruction as seen from point P in Figure E2, plotted onto Waldram diagram

The approach is then to plot the 0.2% sky factor contour in the room, both before and after the new obstruction is erected. Figure E4 shows an example. This contour can be quite time consuming to plot. In most rooms it will lie close to the no-sky line, so it is usually quickest to plot the no-sky line first (Appendix D), then concentrate on points a small distance closer to the windows for the sky factor computation. Interpolation can be used where necessary.

Then find the areas of the room which lie beyond the 0.2% contours 'before' and 'after'. (Note that the 'before' condition should include any obstructions previously on the adjoining site.) According to legal precedent, if more than half a room has a sky factor of less than 0.2%, then the room as a whole is inadequately lit. This is not a hard and fast rule, and in one case a room was deemed inadequately lit even though slightly less than half of it had a sky factor of less than 0.2%. One important factor is whether the proposed new building is one of a number of possible future obstructions which may further reduce the light in the existing room.

As a general guide, though, if after construction of a proposed development more than half a room in an existing building will have a sky factor of less than 0.2%, then any rights of light will probably be infringed. The extent to which an injury of this sort is actionable depends on the difference between the situations 'before' and 'after'. Here there is no clear legal precedent. If the adequately lit (ie, over 0.2% sky factor) part of the room decreases from 50% to 49%, this would be scarcely noticeable and hence an action would be unlikely to succeed. If it decreased from 50% to 30%, this would almost certainly be significant and hence actionable in most habitable rooms. However, between these points it is impossible to give a firm guide on the likely decision of a court. A lot depends on how the room is being used.

Because of these uncertainties, and because the costs of losing a rights-to-light action may run into many thousands of pounds, it is wise for developers to be cautious if there is any possibility that the rights to light of an existing nearby building may be infringed. If, as a result of a proposed development, an existing room will become inadequately lit, as defined here, then the safest course is to reduce the size of the development or increase its distance from the obstructed building.

References to Appendix E

1   **British Standards Institution**. Code of practice for daylighting. *British Standard* BS 8206 Part 2: 1992. London, BSI, 1992.

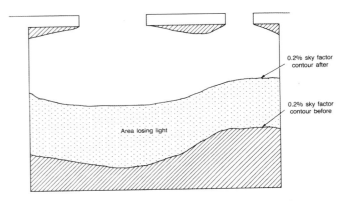

Figure E4   Plan of room showing movement of the 0.2% sky factor contour following erection of an obstruction

Appendix F

Setting alternative target values for skylight access

Sections 2.1, 2.2 and 2.3 give numerical target values in assessing how much light from the sky is blocked by obstructing buildings. These values are purely advisory, and different targets may be used, based on the special requirements of the proposed development or its location. Such alternative targets may be generated from the layout dimensions of existing development, or they may be derived from considering the internal layout and daylighting needs of the proposed development itself. Figure F1 shows one way of doing this. The spacing of the terraced houses is just enough for all the working plane in a room to receive some direct skylight. This is how the 25° angle in the guideline in Section 2.1 was obtained

Section

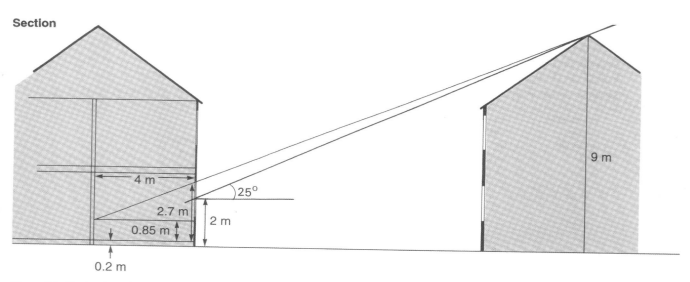

Figure F1 Derivation of spacing target for terraced houses. All points on the working plane in the ground floor room can receive direct skylight

69

Whatever the targets chosen for a particular development, it is important that they should be self-consistent. Table F1 can be used to ensure this. First, choose a limiting obstruction angle (for wide obstructions) from the first column. This is angle γ_1 in Figure F2. The second column expresses this as the ratio (spacing of obstruction s_1):(height above reference point h_1). The third column gives the equivalent vertical sky component at the reference point; this can be used to assess the skylight impact of taller, narrower obstructions. The remaining three columns give the corresponding quantities which can be used to assess the amount of skylight left to reach adjoining development land (Section 1.3). They are derived from the building-to-building angles in the first column, by using the method illustrated in Figure 12 of Section 2.3, which constructs an imaginary 'mirror image' building on the other side of the boundary. Again, all angles and heights are expressed relative to a reference point which, for boundaries, is normally taken to be 2 m above ground level.

Table F1 Equivalent vertical sky components, spacing-to-height ratios, and boundary parameters corresponding to particular obstruction angles between rows of buildings (Heights and angles are usually relative to a point 2 m above the ground — see Figure F2.)

Obstruction angle γ_1 on building (degrees to horizontal)	Equivalent spacing-to-height ratio (s_1/h_1)	Equivalent vertical sky component (%)	Obstruction angle γ_2 at boundary (degrees to horizontal)	Spacing from boundary divided by height (s_2/h_2)	Vertical sky component at boundary
15	3.7	33	28	1.9	26
16	3.5	32	30	1.7	24
18	3.1	31	33	1.5	23
20	2.7	30	36	1.4	21
22	2.5	29	39	1.2	19
24	2.2	28	42	1.1	17
25	2.1	27	43	1.1	17
26	2.1	27	44	1.0	16
28	1.9	26	47	0.93	14
30	1.7	24	49	0.87	13
32	1.6	23	51	0.81	12
34	1.5	22	53	0.75	11
35	1.4	21	54	0.73	10

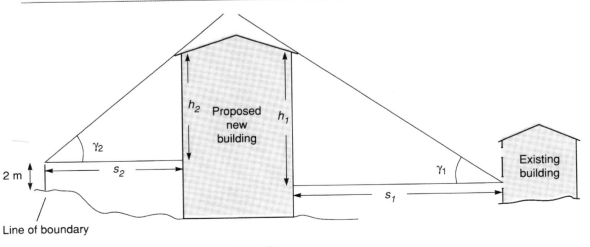

Figure F2 Angles, spacings and heights used in Table F1

Appendix G

Calculation of sun on the ground

A set of 12 transparent indicators, separately available from BRE to accompany this book Ref AP60, can be used to predict the availability of sunlight on the ground at the equinox (March 21). The indicators are for three different latitudes, and for use with four different scales of plan (1:100, 1:200, 1:500 and 1:1250). It is important to use the correct indicator for each location and, especially, for the scale of plan used. If the available plan is drawn to a different scale, the indicator may be modified by changing its height scale. For example, a 1:50 scale indicator may be created by taking the 1:100 indicator and dividing each of the heights by two.

The indicators are calculated for latitudes of 51.5° N (London), 53.5° N (Manchester) and 56° N (Edinburgh/Glasgow). The London indicators may be used for southern England and south Wales; the Manchester ones for northern England, north Wales and the southern half of Northern Ireland. For Scotland and the northern half of Northern Ireland, use the Edinburgh/Glasgow indicators.

Figure G1 illustrates one of the indicators. They work in the same way as the former sunlight indicators which accompanied the 1971 Department of the Environment publication *Sunlight and daylight: planning criteria and design of buildings.* Unlike the indicators in Appendix A, the sun-on-ground indicator can be laid directly onto a plan of that particular scale. In use, the indicator must always be aligned with its south point corresponding to due south on the plan.

The point P at the centre top of the indicator is placed at the reference point on the plan where the sunlight needs to be calculated. The radial lines fanning out from P give the directions of the sun at different times of day. The topmost radial lines (labelled 10° line) give the directions at which the sun's altitude is 10°. Sunlight coming from below this angle is usually discounted, as it is likely to be prevented from reaching the ground by fences, plants or other low-level obstructions.

The horizontal lines running across the indicator give the heights of obstructions which would just stop the sun from shining on point P at each time of day. These heights are relative to the point on the ground. On sloping ground, take care that the height used is the actual height of the top of the obstruction above the particular reference point.

To find the hours of sunshine on March 21 at a particular reference point, lay the correctly scaled indicator on the plan, with point P at the reference point, and the south line pointing due south. For each obstruction, compare its height above the reference point with the height on the corresponding horizontal line on the indicator. If the obstruction is higher, it will block sunlight at the time of day indicated by the hour lines.

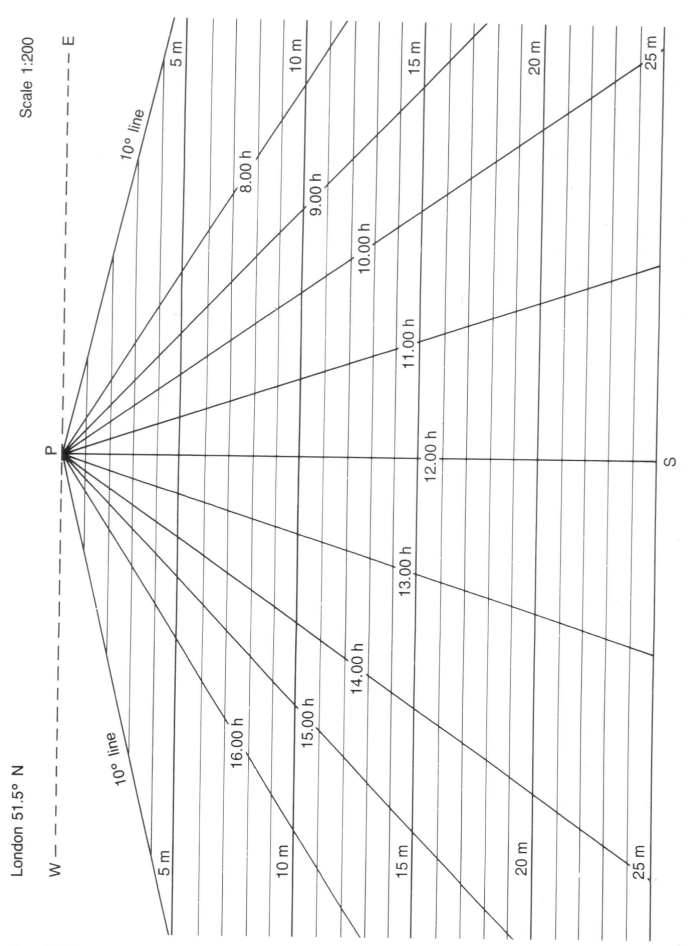

Figure G1 Sun-on-ground indicator for March 21 (for southern England and south Wales, scale 1:200)

Figure G2 illustrates the use of the indicator for this purpose. Building face AB will block sunlight from reaching point P from around 14.15 h onwards. The portion BC of the two-storey extension will block sunlight because, at 8 m high, it is higher than the values on the height scale of the indicator. However, portion CD will not block the sun, because it is lower than the values on the height scale. Similarly, building face EFG starts to obstruct the sun at point F, where the 7 m height line intersects the line EFG on plan. EFG is the side of a building with a sloping roof, and it is also necessary to check whether the ridge blocks any more sunlight. The ridge stops blocking the sun at point I on the 9 m height line. This corresponds to just before 08.00 h. As face EF blocks the sun until just after 08.00 h, in this case the eaves of the building block more sunlight than the ridge. This type of double check also needs to be done when a taller building lies behind a smaller one.

The net result is that point P receives more than $4\frac{1}{2}$ hours of sunlight, from just after 08.00 h until 12.45 h.

The indicator can also be used to find those areas of ground which cannot receive any sunlight at all on March 21 (see Section 3.3). It is possible to use trial and error for this, repeating the procedure for a grid of points on the open space. But, for most simple cases the calculation can be shortened considerably.

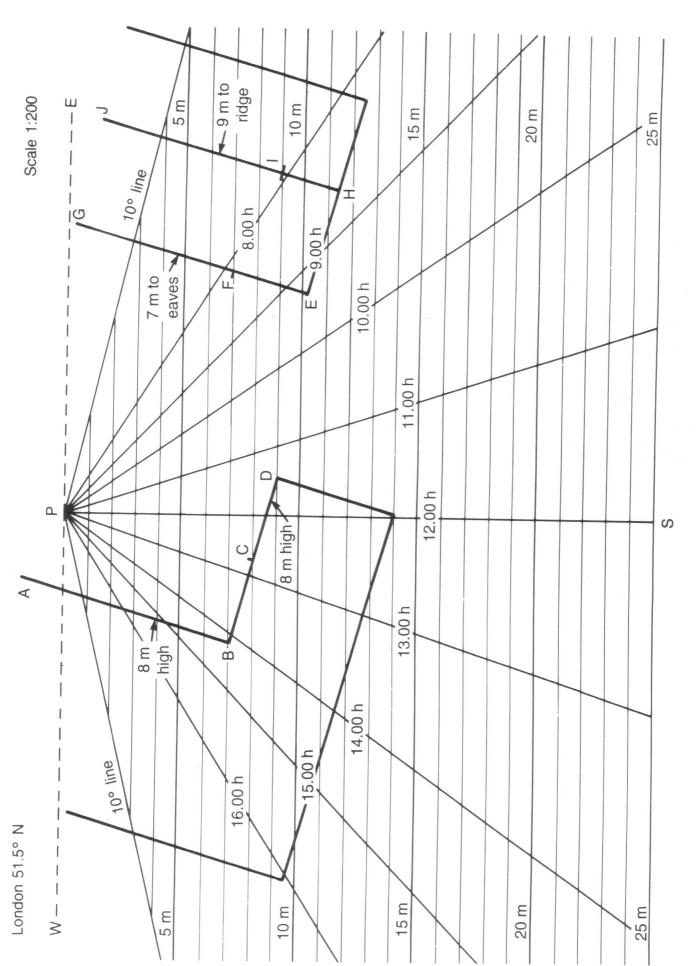

Scale 1:200

London 51.5° N

Figure G2 Sun-on-ground indicator superimposed on an example situation

As Section 3.3 pointed out, if an area cannot receive direct sun on March 21, this probably comes about in one of two ways. The first is illustrated in Figure G3, and the second in Figures G4 and G5.

In the first situation, a long side of a building faces within 13° of due north. The area of ground directly adjoining the building face is shaded all day long at the equinox. In Figure G3, AD is the side of a long, single-storey block with its roof 3 m above ground level. It could be a school classroom block, or a hotel annexe. This side AD faces 3° from due north. The area permanently shaded is found in the following way. First draw along the 10° line AB from the southernmost corner of the building face A. Points north of this line will receive some sunlight around the building in the early morning. Move the indicator along this line until the corner A intersects the horizontal height line corresponding to the height of the building (in this case the 3 m line). Figure G3 shows the indicator in this position, at point B. Mark this point through the hole in the acetate sheet. The boundary of the area of shade then runs parallel to the building face, along line BC. Points north of this line receive some sunlight over the top of the building on March 21. At point C, this line BC intersects the 10° line drawn from the other corner of the building face D. Points north of line CD will receive some sunlight in the late afternoon around the western side of the building. Thus, the area permanently shaded on March 21 is ABCD.

Note that if the building face AD had been less wide, or taller, the area shaded all day would have been a triangle bounded by the two 10° lines drawn from the corners of the building, A and D. (It should be emphasised that the 10° line represents 10° in solar altitude, not solar azimuth: it is in fact 13° from due east or west in London, 15° from due east or west in Edinburgh.)

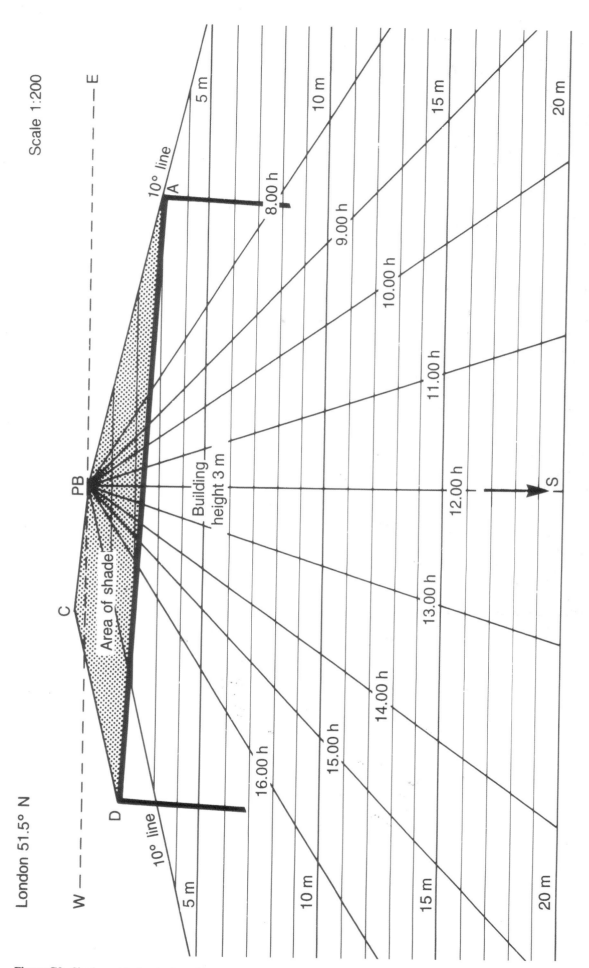

London 51.5° N

Scale 1:200

Figure G3 Shade cast in front of an almost north-facing building. Area ABCD cannot receive sunlight at any time on March 21

In the second situation, permanently shaded areas of ground occur when buildings are set at an angle to each other, to create an enclosed or partly enclosed space which faces north. Figure G4 shows a typical example: a row of terraced houses ABFC, with a single-storey extension BDEF to its north–east side.
(In this example the extension is unusually long, to show the shading clearly.) The area of ground by point B will be in permanent shade on March 21, as it cannot 'see' any of the southern half of the sky. To work out the extent of this area, first find the points T and R, on each obstructing face, which just receive sunlight above point B on the perpendicular face. Figure G5 shows the indicator at point R; point B is on the 7 m height line, which is the height of the eaves at B. With a sloping roof, it should be checked whether the ridge blocks more sunlight; here it does not, because point B' is beyond the 9 m height line. If the ridge had been much higher, then point R would have ended up farther from the main terrace. Note also that point R has been marked on an imaginary continuation of the extension face BD, because the distance BR is, in this example, more than the length of the extension.

Point T on the other face is marked in the same way. This time point B is on the 3 m height line, because that is the height of the side of the extension which just stops the sunlight from reaching T. Then the border of the no-sun area is the line TR. Points north of this line will receive sunlight either above the extension, or the main building, or both. If the two obstructing 'arms' of the building had been the same height, line TR would have run exactly east–west.

In this particular case, since point R lay on an imaginary continuation of the extension face BD, it will in fact receive sunlight from the east around the end of the extension, as Figure G5 shows. Allow for this by drawing along the 10° line (see previous example), from point Q on the line TR to the end of the extension face at D. Points in the triangle QRD will receive early morning sunlight around the end of the extension. Thus, the part of the garden which can receive no sunlight on March 21 is the shaded area BTQD on Figure G4.

This method applies strictly only if the two arms of the building are roughly the same height, or if the higher part of the building extends some way beyond the lower part, as in Figure G4. If the main building had ended between B and F instead of at C, then some of the area BTQD could have received sunlight above the extension and around the eastern side of the main building. In most practical situations the difference in area is small. The main exception to this occurs if the main building actually ends at point B, level with the face at right angles to it. An example is shown in Figure G6.

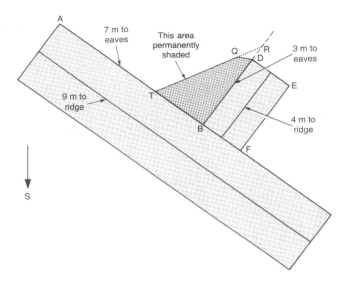

Figure G4 Sun-on-ground calculation for an extension to a long terrace

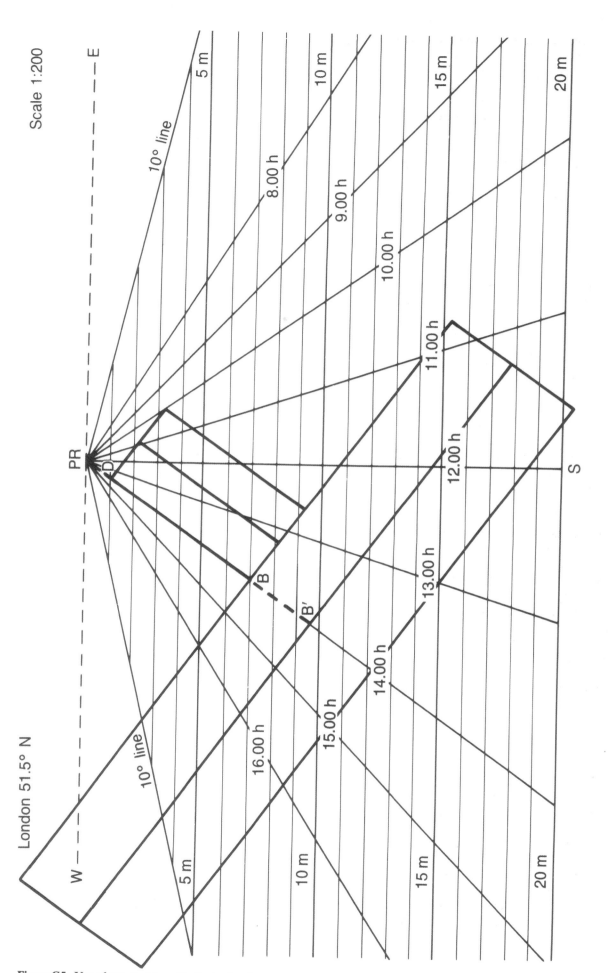

Figure G5 Use of sun-on-ground indicator for the example shown in Figure G4

79

In Figure G6, the single-storey extension DEFG is at the side of the house(s). Point T, on the main face of the house, is found in the same way as for Figure G4. But this time the border of the sunless area goes directly eastwards from T, to strike the extension at U. Point B is the critical part of the obstruction, for points north of line TU will receive direct sunlight over the extension near point B, whatever the height of the main building. Therefore, triangle BTU will be the only area permanently shaded on March 21.

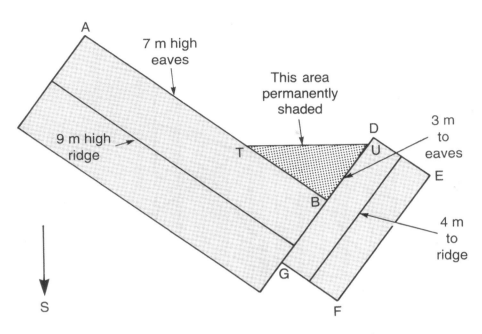

Figure G6 Sun-on-ground calculation for an extension at the side of a house

Figure G7 illustrates another situation, where the obstructing buildings are detached from each other. It is the same as Figure G4, except that BDEF is now a separate building rather than an extension. The area of shade is found in exactly the same way as for Figure G4, except that the area which can receive sunlight through the gap between the buildings needs to be subtracted. This is (approximately) area B'BUT in Figure G7. Thus the area permanently shaded on March 21 is approximately BUQD. In fact, the actual area may be slightly smaller than this, because some points near UB may receive light through the gap and over the top of the main building, but in practice this makes little difference.

It is worth emphasising that a point is not necessarily well sunlit just because it is outside the 'permanently shaded' region. For example, in Figure G7 the area B'BUT can receive some sunlight on March 21, and indeed throughout the year, but it will be for only a few minutes each day. If more than a minimum amount of sunlight is required at a particular place, then use the sunpath indicator in Appendix A, or at least calculate the duration of sunlight on March 21 using the method described at the start of Appendix G.

Note that all the examples have assumed level ground next to the building. If the ground slopes significantly, a trial-and-error approach with the indicator may be necessary. If the ground is terraced, stepping down from one level to another, the calculation may be repeated for each level and the results juxtaposed on plan.

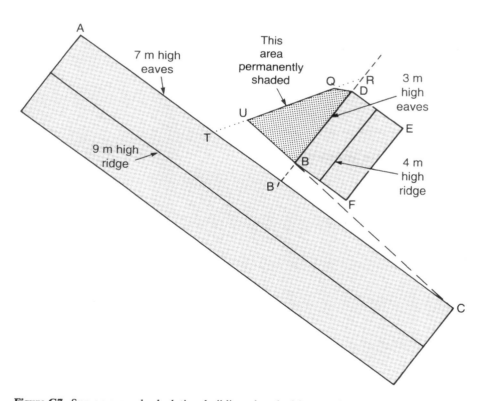

Figure G7 Sun-on-ground calculation; buildings detached from each other

A third use of the sun-on-ground indicator is to plot the course of the shadows cast by a building throughout the day. This is especially valuable for a large development which may overshadow a number of neighbouring properties. Again, a plan of the scale marked on the indicator is required, but this time the indicator is used in reverse. Place point P on the indicator at each point on the roofline of the building in turn. The south (noon) point on the indicator must point due north, because shadows will be cast in this direction at midday. Then the direction of the shadow cast by the point on the roofline at a particular time of day is the direction of the hour line. The shadow ends where this hour line intersects the height line corresponding to the height of the point on the roof line above the ground at the end of the shadow.

Figure G8 gives the outline plan and elevation of a simple example; a medium-sized office block with an atrium. Figure G9 illustrates the use of the indicator to determine the shadow cast by point C at the apex of the atrium roof, at 10.00 h on March 21. Point P on the indicator is placed on point C on the plan. As point C is 16 m above the ground, its shadow falls at point C' where the 10.00h line intersects the 16 m height line. Point C' is marked on the plan, and the procedure then repeated for points A, B, D and E on the roof line. The resulting shadow is found by simply joining the dots (Figure G9). Note that the shadows of vertical edges should be parallel to the hour line. The method can be repeated for different times to build up a diurnal profile of shadowing at the equinox; each of the points A', B', C', D' and E' will move in a straight line from west to east through the day.

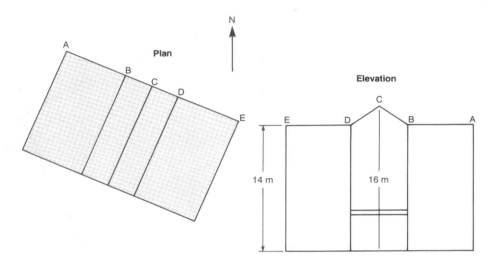

Figure G8   Outline plan and northern elevation of a shadow plotting example building

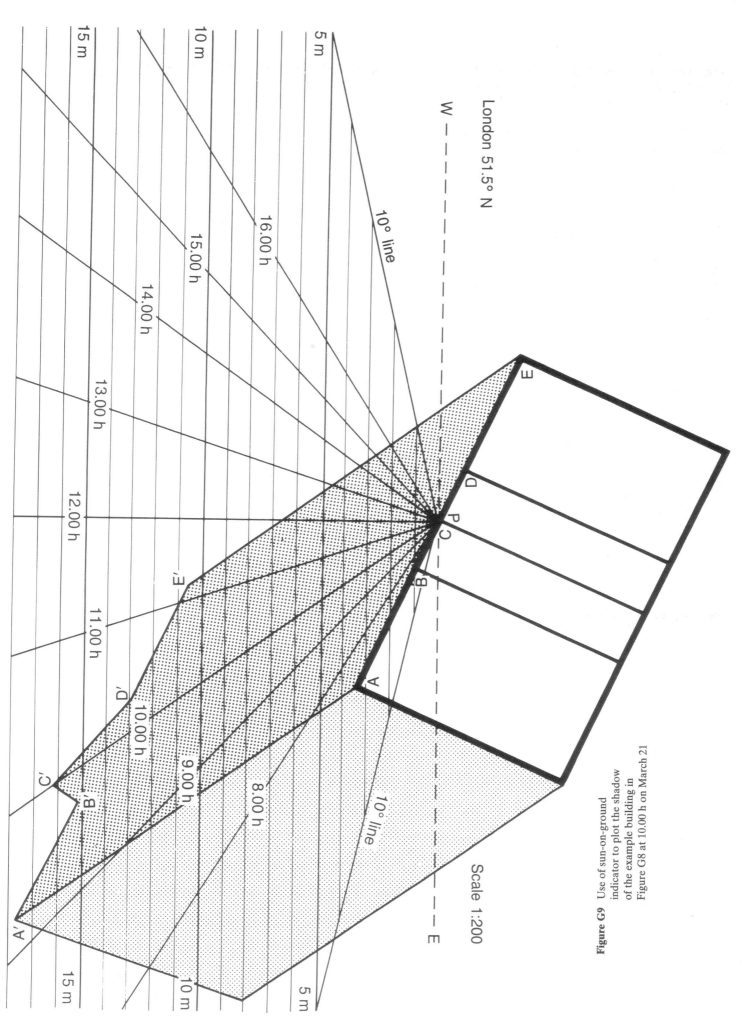

Figure G9 Use of sun-on-ground indicator to plot the shadow of the example building in Figure G8 at 10.00 h on March 21

London 51.5° N

Scale 1:200

Appendix H

Definitions

Average daylight factor
Ratio of total daylight flux incident on the working plane to the area of the working plane, expressed as a percentage of the outdoor illuminance on a horizontal plane due to an unobstructed CIE Standard Overcast Sky (CIE = Commission Internationale de l'Eclairage — the International Commission on Illumination)

CIE Standard Overcast Sky
A completely overcast sky, for which the ratio of its luminance L_γ at an angle of elevation γ above the horizontal to the luminance L_z at the zenith is given by

$$L_\gamma = \frac{L_z (1 + 2 \sin\gamma)}{3}$$

Daylight, natural light
Combined skylight and sunlight

No-sky line
The outline on the working plane of the area from which no sky can be seen

Obstruction angle
The angular altitude of the top of an obstruction above the horizontal, measured from a reference point in a vertical plane in a section perpendicular to the vertical plane

Probable sunlight hours
The long-term average of the total number of hours during a year in which direct sunlight reaches the unobstructed ground (when clouds are taken into account)

Sky factor
Ratio of the parts of illuminance at a point on a given plane that would be received directly through unglazed openings from a sky of uniform luminance, to illuminance on a horizontal plane due to an unobstructed hemisphere of this sky

Vertical sky component
Ratio of that part of illuminance, at a point on a given vertical plane, that is received directly from a CIE Standard Overcast Sky, to illuminance on a horizontal plane due to an unobstructed hemisphere of this sky

Working plane
Horizontal, vertical or inclined plane in which a visual task lies. Normally the working plane may be taken to be horizontal, 0.85 m above the floor in houses and factories, 0.7 m above the floor in offices

Index